Palgrave Studies in Creativity and Culture

Series Editors
Vlad Petre Glăveanu
Department of Psychology
Webster University
Geneva, Switzerland

Brady Wagoner
Communication and Psychology
Aalborg University
Aalborg, Denmark

Both creativity and culture are areas that have experienced a rapid growth in interest in recent years. Moreover, there is a growing interest today in understanding creativity as a socio-cultural phenomenon and culture as a transformative, dynamic process. Creativity has traditionally been considered an exceptional quality that only a few people (truly) possess, a cognitive or personality trait 'residing' inside the mind of the creative individual. Conversely, culture has often been seen as 'outside' the person and described as a set of 'things' such as norms, beliefs, values, objects, and so on. The current literature shows a trend towards a different understanding, which recognises the psycho-socio-cultural nature of creative expression and the creative quality of appropriating and participating in culture. Our new, interdisciplinary series Palgrave Studies in Creativity and Culture intends to advance our knowledge of both creativity and cultural studies from the forefront of theory and research within the emerging cultural psychology of creativity, and the intersection between psychology, anthropology, sociology, education, business, and cultural studies. Palgrave Studies in Creativity and Culture is accepting proposals for monographs, Palgrave Pivots and edited collections that bring together creativity and culture. The series has a broader focus than simply the cultural approach to creativity, and is unified by a basic set of premises about creativity and cultural phenomena.

More information about this series at
http://www.palgrave.com/gp/series/14640

Todd Lubart
Editor

The Creative Process

Perspectives from Multiple Domains

Editor
Todd Lubart
Paris Descartes University
Paris, France

Palgrave Studies in Creativity and Culture
ISBN 978-1-137-50562-0   ISBN 978-1-137-50563-7   (eBook)
https://doi.org/10.1057/978-1-137-50563-7

Library of Congress Control Number: 2018943851

This Palgrave Macmillan imprint is published by the registered company Springer Nature Limited
The registered company address is: The Campus, 4 Crinan Street, London, N1 9XW, United Kingdom

Preface

Although creativity is increasingly viewed as a key human ability that contributes to individual's personal development, everyday life problem solving, professional accomplishment, and societal growth, the process that underlies the production of creative work remains somewhat mysterious. There have been many case studies of artists, writers, inventors, composers, and other professionals. Based on these studies, some theories of the creative process have been proposed. For example, one theory suggests that there is a preparation phase, then incubation, illumination, and finally, elaboration and refinement of ideas. This simple model does not, however, correspond to some detailed accounts of creative work nor does it offer many insights into the differences between individuals who produce ideas that range widely in originality. Research that compares, in detail, how people who engage in creative work as part of their daily jobs get their novel ideas is surprisingly rare.

This edited book seeks to advance work on the creative process by presenting work from a diverse set of creative fields—visual arts, writing, science and engineering, design, and music. For each field, there are two chapters: One chapter is informed largely by accounts of the creative process by those in the field, be they artists, writers, scientists and engineers, designers, or musicians. The authors of each chapter are themselves engaged in their respective fields, and offer insights from an "insider's" view, which are combined with perspectives developed in the academic literature. The other chapter in each section presents results of empirical research in which people completed creative projects in the respective field and their activity was traced and analyzed. The research presented in these latter chapters stems mainly from a four-year research project entitled "CREAPRO", which sought to compare and contrast the creative process in diverse professional domains of creative activity. In each domain, a common methodology was used consisting of structured interviews of professionals working in the field (artists, writers, scientists, designers, and composers). These interviews were followed by observational studies with students in the fields who led creative projects and allowed their activity to be traced over several sessions. Finally, some process-training studies were conducted to see the extent to which the creative process could be influenced by learning activities. This work was conducted in France and supported by the French national research agency.

Paris, France   Todd Lubart

Note

This project was supported by the French National Research Agency (Agence Nationale de Recherche, grant no. ANR-08-CREA-038, CREAPRO) on the empirical study of the creative process.

Contents

Notes on Contributors

Sergio Agnoli is senior researcher at the Marconi Institute for Creativity (MIC). He has held teaching appointments at the University of Bologna, University of Padova, and University of Ferrara. His research interests include: cognitive and emotional substrates of creative thinking; creative thinking process; emotional development and emotional intelligence; and psycho-physiology of emotions. In these fields, Sergio Agnoli has published many contributions in peer-reviewed international journals and conferences and he established collaborations with several research groups and universities.

Baptiste Barbot is Assistant Professor in Psychology at Pace University, Department of Psychology and Yale University, Child Study Center, USA. He is Associate Editor of *Psychology of Aesthetics, Creativity and the Arts*, and *New Directions for Child and Adolescent Development*. His work focuses on creativity measurement and development in adolescence.

Nathalie Bonnardel is a Full Professor of Cognitive Psychology and Ergonomics at Aix-Marseille University. She conducts her research at the Psychology of Cognition, Language and Emotion (PsyCLE) laboratory, where she heads the Cognition, Emotion and Expertise research team.

She is also the Director of the Master's Degree in Ergonomics: Human Factors and Information System Engineering. Her twin research aims are to improve current understanding of creative design activities and to support these activities. Her approach consists in using results obtained in the fields of cognitive psychology and ergonomics to develop methods and computational systems that promote users' activities.

Marion Botella is assistant professor in Differential Psychology at the Paris Descartes University. After defending her thesis describing how emotions are involved in the artistic creative process, she was postdoctoral researcher at the Université Catholique de Louvain (Belgium) where she examined the impact of creativity on mood. Since 2013, she is conducting her research at the LATI lab (Laboratoire Adaptations Travail Individu), Paris Descartes University. Her research focuses on (1) the creative process in various domains (as art, science, design, etc), (2) alexithymia and affective traits of creators, and (3) the development and construction of scales.

Samira Bourgeois-Bougrine earned her Ph.D. in ergonomics and human factor engineering. Her initial research focused on human-machine interactions, ergonomic analyses of behavior, and best practices in the aviation agency in terms of pilots and air traffic controllers. Samira Bourgeois-Bougrine went on to focus her research on analyses of best practices in the creative process of engineers and writers, based on in-depth analyses of responses to interviews and activity traces. Her current work focuses on ways to enhance creativity in educational settings, through pedagogies that seek to improve specific aspects of the creative process.

Stéphanie Buisine is a Professor in Innovation at EI.CESI school of engineering in Paris-Nanterre (Laboratoire d'Innovation Numérique pour les Entreprises, les Apprentissages et la Compétitivité des Territoires), and associate member of Paris Descartes University (Laboratoire Adaptations Travail Individu). Her multidisciplinary background includes Psychology, Ergonomics, Human-Computer Interaction and Industrial Engineering. Her research focuses on the methodological, technological, and organizational factors of innovation, with a special emphasis on need-seeker strategy, prospective methods, technology-supported creativity, and innovation culture.

Giovanni Emanuele Corazza is a Full Professor at the Alma Mater Studiorum-University of Bologna, Member of the Alma Mater Board of Directors, founder of the Marconi Institute for Creativity, Member of the Marconi Society Board of Directors, and President of the Scientific Committee of the Fondazione Guglielmo Marconi. He is the originator of the dynamic definition of creativity. He is author of more than 300 papers, and has been the General Chairman of the IEEE ISSSTA 2008, ASMS 2004–2012 Conferences, and MIC Conference 2013–2016. His research interests are in creativity and innovation, 5G systems, navigation, and positioning.

Benjamin Frantz holds a master degree in psychology and worked as research engineer at the LATI laboratory and the SONY Computer Science Laboratory. He participated in several research projects on creativity and its measurement. He is currently a stone carver and artist.

Pierre-Yves Gilles is a Full Professor of Differential Psychology at Aix-Marseille University and conducts his research at the PsyCLE laboratory. He is also Dean of the University's Faculty of Humanities, where he sees his primary role as guiding students in their choice of studies and subsequent integration into the world of work—his twin research topics. He is particularly interested in how students adapt to university life, in terms of individual differences in creativity, aptitudes, emotions and motivation.

Vlad Petre Glăveanu is Associate Professor and Head of the Department of Psychology and Counselling at Webster University Geneva, Switzerland, Associate Professor II at the Center for the Science of Learning and Technology (SLATE) at Bergen University, Norway, and Director of the Webster Center for Creativity and Innovation (WCCI). He completed his Ph.D. in Social Psychology at the London School of Economics. Editor of the *Palgrave Handbook of Creativity and Culture* (2016) and the *Oxford Creativity Reader (2018)*, Vlad received in 2018 the Berlyne Award for outstanding early career contributions to the study of aesthetics, creativity, and the arts.

Todd Lubart is Professor of Psychology at Paris Descartes University. He is the author of numerous books, articles, and book chapters about creativity. Also, he has co-authored a measure of creative potential in children (EPoC: Evaluation of Potential Creativity). Todd Lubart received awards for his work on creativity including the American Psychological Association, World Council of Gifted and Talented, and International Center for Innovation in Education. He directs a research laboratory (LATI, Laboratoire Adaptations Travail Individu) that focuses on individual differences and applied psychology, with a focus on creativity. Todd Lubart has been responsible for several large-scale research grants, including the "CREAPRO" project on which this book is based.

Daniel Martín is a musician and software engineer. He holds a Master in Sound and Music Computing (UPF—Barcelona) and has studied jazz saxophone, classical guitar, and music composition with new technologies. During his work at Sony CSL, he has designed a lead sheet database which at the time of writing contains more than 15,000 lead sheets in the style of jazz, bossa-nova, and popular music. He also has developed a web-based editor for lead sheet composition in which users can provide feedback to each other.

Sylvain Mazon is a psychologist and ergonomist. After earning a Master's Degree in Ergonomics: Human Factors and Information System Engineering at Aix-Marseille University, he became a freelance consultant/trainer. He is the founding director of Soluce Ergonomie (Alès, France) and teaches at the Conservatoire National des Arts et des Métiers (CNAM) in Montpellier (France).

Julien Nelson is assistant professor in ergonomics and human-computer interaction at Paris Descartes University, Laboratoire Adaptations Travail Individu. His research interests focus on human factors methodologies to support the design of interfaces using emerging technologies, and the use of Virtual Reality technology to support creative activities.

François Pachet is currently research director at Spotify and former director of the SONY Computer Science Laboratory Paris, where he led the music research team. He received his Ph.D. and Habilitation degrees from Université Pierre et Marie Curie (UPMC). He is a Civil Engineer (Ecole des Ponts and Chaussées) and was Assistant Professor in Artificial Intelligence at UPMC until 1997. He joined the Sony Computer Science Laboratory in 1997 and created the music team to conduct research on interactive music listening, composition and performance. He has been elected ECCAI Fellow in 2014. He is also an accomplished musician (guitar, composition) and has published two music albums (in jazz and pop) as composer and performer. His current goal, funded by an ERC Advanced Grant, is to build a new generation of tools to assist music creation. He released the first mainstream music album composed with AI, Hello World, in 2018.

Jane Piirto is Trustees' Distinguished Professor at Ashland University in Ohio, USA. She is the author of 21 books, among them *Organic Creativity in the Classroom*; *Creativity for 21st Century Skills*, *Understanding Creativity*; *My Teeming Brain: Understanding Creative Writers*, *Understanding Those Who Create* (2 editions); and *Talented Children and Adults: Their Development and Education* (3 editions); Literary books are available on Kindle and Nook. She has also published poems, studies, articles, and chapters in edited books. She received the Mensa Lifetime Achievement Award, the National Association for Gifted Children Distinguished Scholar Award, and the Torrance Creativity Award, among others.

Ivan Toulouse is Professor at the art department in Rennes 2 University in France. He lives and works as an artist in Paris and his research is based on art practice, as are the Ph.D.s he supervises. Research in art is a reflective attempt to make explicit what has occurred pragmatically in the creative process, by experimenting it, comparing it with other experiences, and problematizing it theoretically. Art is a laboratory to approach how human mind really works. This is the subject of his book, Clair-obscur. Pour une pensée de la création, Paris, Ed. L'Harmattan, Coll. «Eurêka & Cie», 2012.

Peter R. Webster is a Scholar-in-Residence at the Thornton School of Music at the University of California in Los Angeles, USA and Emeritus Professor at Northwestern University in Evanston, Illinois, USA. He has over 90 publications including books and book chapters on music creativity in children, music technology, and others related to music teaching and learning.

Alicja Wojtczuk is a psychologist and ergonomist. After earning a Master's Degree in Ergonomics: Human Factors and Information System Engineering at Aix-Marseille University, she completed a Ph.D. thesis in Cognitive Ergonomics, supervised by Prof. Nathalie Bonnardel. Her research project focused on creativity within the design process. She currently works as a user experience researcher at MyScript (France), studying the handwriting process.

List of Figures

List of Tables

1

Introduction

Todd Lubart

Creativity is an important twenty-first century skill that involves producing new, original work that has value and meaning in its' context. Creativity is considered to be a key ability that will promote well-being for individuals, organizations, and the society at large. It is recognized as important in surveys of educators (Berland, 2013), business leaders (Berman & Korsten, 2010) and international organizations such as the Organization for Economic Cooperation and Development (OECD) and the World Economic Forum.

In an overview of work on creativity, Lubart (2017) proposed a "7 Cs" perspective as an organizing framework. The seven C's are: Creators, Creating, Collaboration, Context, Creations, Consumption and Curricula. The term *Creators* refers to studies of characteristics of creative people. *Creating* concerns the working process. *Collaboration* signifies interactions with colleagues and significant others. *Context* denotes the physical and social environment. *Creations* are the outputs or works that result. *Consumption* is the uptake of creations by the public, and

T. Lubart (✉)
Paris Descartes University, Paris, France

© The Author(s) 2018
T. Lubart (ed.), *The Creative Process*, Palgrave Studies in Creativity and Culture,
https://doi.org/10.1057/978-1-137-50563-7_1

the term *Curricula* refers to training and developing creativity. These different topics are interrelated. For example, the characteristics of creators, those who engage in the creative thinking, include specific cognitive abilities such as associative ability, metaphorical thinking, evaluative thinking and many others as well as knowledge. The way these different cognitive abilities and content knowledge enter into the chain of thoughts and actions that eventually lead to a production, such as an artistic work, is the process. It is impossible to examine the *process* without considering the resources (abilities, traits, contextual elements) that enter into play. Thus, research on the creative process is naturally situated in a specific context.

The field of creativity research has been growing steadily and within the field there is exciting work on all 7 Cs. However, not all Cs have received equal coverage. A simple search reveals, for example, that there has been a relatively greater number of studies on the characteristics of creative people. On the general topic of the creative process, there has been increasing attention in recent years. Based on the PsycINFO database, for example, searching for the term "creative process" in the title of the article, book or chapter, there were 9 abstracts from the start of the database from 1927 to 1947, 33 abstracts from 1948 to 1967, 164 abstracts from 1968 to 1987, 251 abstracts from 1988 to 2007, and 260 abstracts from 2008 to today, which is a period of only 10 years.

The creative process is, in many ways, the hub of the 7 Cs. Creators' abilities and traits come into play in the process, Collaboration refers to interactions with other people or with objects which have a process component, Context is where the process unfolds. Creations are the trace that results from the process. Consumption is the uptake of creative productions and itself involves a diffusion process. Finally, the last C, Curricula, refers to the development of creativity, which is a process—and often techniques to help structure the creative process are taught.

These considerations led to this volume which explores the creative process in several fields of endeavor. The domains chosen—visual arts, design, science and engineering, writing, and musical composition—represent diverse "creative" endeavors. People who work as artists, designers, scientists and engineers, writers and composers engage in the creative process as part of their core activities on the job. It should be noted that the creative process is not an exclusive activity that concerns only "creative" fields. In fact, many jobs may benefit from creativity, at least some of the time. This is the case, for example, of managers, teachers, lawyers, and a vast number of professions. In addition, the creative process is present in diverse everyday life activities, such as solving problems among friends, inventing a new dish for dinner, or engaging in recreational craft activities.

The Creative Process—What Are We Talking About?

The creative process can be defined as a sequence of thoughts and actions that comprise the production of work that is original and valuable. This initial definition codes many implicit facets of the "creative process" concept which deserve further explanation. A "sequence" signifies that there is a chain of events, which unrolls over time, with a beginning and, potentially, an end. This sequence may be non-linear, it may be characterized by steps or phases, or activities that come into play at certain moments in the chain of events. "Thoughts and actions" refer to both internal and external operations that contribute to the emerging production. "Production of work" denotes an outcome which may be a tangible or intangible thing that is expressed in some form. The work may be expressed visually, verbally, acoustically, mathematically, kinetically or in other ways and may have a stable, permanent or a more ephemeral nature.

The reference to "original" is important because this corresponds to the standard definition of creativity (Runco & Jaeger, 2012). The work must be different from the prior existing productions that the individual creator has made. The idea of novelty is present here as a hallmark of creativity. This novelty must be present, at least, for the person, or people, involved in the process. However, the extent to which the production is original in a wider social context (local comparison, or in the field of endeavor), the more the work will be socially deemed creative. It is possible for a person to engage in a process that leads to original work for the individual, but it is not original for other people. It is still relevant to talk about the creative process in this case.

The use of the term "valuable" reflects as well the standard definition of creativity. Creative work corresponds to a need, a goal, or fits within its' context. This value may be appreciated by the creator and/or by the audience. It is useful to highlight the concept of "value" because this allows original, meaningless productions to be excluded from the realm of creativity.

To complete this analysis, it is useful to distinguish the *creative process* from the *process of creation*. The process of creation brings something into being. However, it may be a repetitive act that creates more of an existing item, such as standardized sequence that yields a standardized result. There is creation, but it is not necessarily original. The creative process requires novelty and value. It is a process of creation that leads to original work. This nuance focuses our attention on the "creativity" angle and allows run-of-the-mill processes that produce run-of-the-mill products to be set apart.

The Creative Process—A Little History

Act 1

Since ancient times, the creative process has been the source of interest and speculation. The "muse" as evoked in historical Greek texts involved a divine source which inspired mortals to express new ideas. The inspiration was divine but the delivery process involved people who served as scribes or oracles. The creative process was conceived as a receptive and then expressive action. The individual creator was a go-between, a medium through which ideas were expressed. The way this happens is rather mysterious, and the creative process was indeed a black-box experience. This is epitomized by Archimedes' eureka experience, which seemed to happen in a flash (see Koestler, 1964; Weisberg, 1993).

Act 2

Fast forward to the Renaissance; the growth of the arts and the sciences was accompanied by a progressive shift from the creative process stemming from an external source of inspiration to an internal one. Some individuals became known for their creative ability, standing out among peer artists, musicians, writers, scientists and engineers. Creative industries began to emerge.

Speculations on the way that outstanding individuals got their ideas led to introspective accounts of the creative experience. For example, at the end of the nineteenth century, Hermann von Helmholtz, the physicist and physiologist, described how after investigating a problem thoroughly, "happy ideas came unexpectedly without effort, like an inspiration" (cited in Wallas, 1926, p. 80). Ideas did not come if he was tired or at his working table, but rather when he was taking a break such as a walk outside (Wallas, 1926).

Henri Poincaré, a French Mathematician, provides an excellent example of this in his description of mathematical invention. He was working on a certain class of functions, Fuchsian functions, that were not well understood at the time. His first step was to work on the problem in a deliberate way, at his desk. He "tried a great number of combinations and reached no results" (p. 26). One evening, he could not sleep and Poincaré reports that "Ideas rose in crowds; I felt them collide until pairs interlocked, so to speak, making a stable combination....I had established the existence of a class of Fuchsian functions" (p. 26). Poincaré wrote down his ideas, elaborated on them, and using an "analogy with elliptic functions," developed his ideas on Fuchsian functions in more details.

As Poincaré had numerous activities, he was solicited to take part on a geological survey near Caen, in the France region of Normandy. He put his work on Fuchsian functions aside temporarily. However, at one point as he got on a bus in Caen, an idea came to him "without anything...seeming to have paved the way for it" (p. 26). Fuchsian functions were in some ways similar to those in non-Euclidean geometry. Once back, Poincaré checked his new insights and then started working on other mathematical problems. The summer period led him to travel again, this time near the sea at Etretat. As Poincaré took a walk, another idea came with "brevity, suddenness, and immediate certainty" (p. 26). Poincaré went back to Fuchsian functions, applied his new idea but then discovered further difficulties. The saga does not end here, because Poincaré had to go for military service. He was "very occupied", but further ideas on his mathematical work "suddenly appeared" during his military activities. Later, back from his service, "he had all the elements and had only to arrange them and put them together" (p. 26). He was able, at that point, to finish directly the mathematical proof.

Poincaré (1908/1985) in his well-known introspective account of the creative process proposes that the creative process starts with conscious work on a problem. Then a period of unconscious work follows, and results in an insight, a "sudden illumination." The cycle back to conscious work occurs, called the verification stage, "to put in shape the results of this inspiration," to explore the consequences and to formalize ideas (p. 27).

Graham Wallas proposed the four-stage model of the creative process: (a) preparation, (b) incubation, (c) illumination, and (d) verification. Preparation may include problem exploration and definition. It is a conscious phase that may extend over a long period, including earlier educational experiences that all prepare for work on a problem. Incubation occurs, in contrast, in the unconscious or at the fringe of consciousness. Perhaps taking a break consciously does not stop problem processing at a deeper level of thought. At least, this is one possible explanation of the incubation process. Associative thinking may be active during incubation, as well as during the preparation phase. Poincaré claimed that his mind monitored ideas, unconsciously, and served as a "delicate sieve" that led only the most interesting, and mathematically aesthetic ideas pass into consciousness. It is interesting to note here a link with the process of forming idea combinations through random or chance-based processes was advanced by Campbell (1960) and further developed by Simonton (2011), in the blind variation—selective retention model. This passage, which may be experienced as an abrupt event, is "illumination". For Wallas, illumination was often announced by a feeling of knowing, an "intimation," at the "fringe" of consciousness (p. 97). During verification, ideas are refined, developed and formalized. Although the four-stages in Wallas' model appear rather orderly, and sequential, it is best to consider them as process states that may occur in cyclical, or varied patterns. It is interesting to read early work by Catherine Patrick (1935, 1937, 1938), who observed artists, writers, and laypeople engaged in creative work, using an observational grid based on Wallas' concepts.

Act 3

There was increasing attention to intelligence and problem solving in the mid twentieth century. In this context, Joy P. Guilford, as president of the American Psychological Association, presented his work on creativity, particularly divergent thinking. Concerning the popular four-stage description of the creative process, Guilford (1950) noted that "such an analysis is very superficial from the psychological point of view. It is more dramatic than it is suggestive of testable hypotheses. It tells us almost nothing about the mental operations that actually occur" (p. 451). Furthermore, "it is not incubation itself that we find of great interest. It is the nature of the processes that occur during the latent period of incubation, as well as before it and after it" (p. 451). Problem-solving abilities involved in creativity may be sensitivity to problems, the capacity to produce many ideas (fluency), flexibility, restructuring ability, complex information processing, and evaluation skills. Guilford's (1967) Structure of Intellect model indicated basic cognitive operations that were applied to content domains (verbal, visual, acoustic, ...). Divergent thinking, convergent thinking and evaluation are particularly important for creativity.

Divergent thinking refers to idea search in multiple directions, an expansive form of cognition, which is inherently an exploration of a thought space. The result is several ideas, which tend to represent diverse angles on the topic, thus showing flexibility. Convergent thinking refers to bringing together various elements. It involves a synthesis. The extent to which this synthesis is constructed and not the result of algorithmic procedures, the convergent operations are relevant to creativity. Thus, it is useful to distinguish creativity-relevant convergent thinking from "getting the right answer" in standard cognition. Evaluative thinking, assessing the strengths and weaknesses, the value of ideas, is a third cognitive ability involved in all problem-solving, including creative thinking. These specific processes enter in the creative workflow, in cyclical ways, with divergence, convergence and evaluation in various orders and intensity.

Guilford's work, in particular his focus on divergent thinking as a particular aspect of creative thinking, led to two practical lines of work that impact visions of the creative process today. One line of work led to psychometric creativity tests. The second led to creativity techniques.

In the first line of work, Guilford's initial measures of divergent thinking developed into well-known "creativity" tests. The prototypical task, used by Guilford, was the alternative uses test which requires the generation of as many different uses for a common object (such as a brick) as possible. It is scored for the number of ideas (fluency), the diversity of ideas (flexibility) and the originality of ideas (originality). Torrance developed these measures into the Torrance Tests of Creative Thinking (TTCT), which have been used widely since the 1960s. The TTCT measure mainly divergent thinking in verbal and graphic production tasks. For example, individuals are asked to generate as many unusual uses as possible for a box, as many ideas as possible to improve a toy, as many drawings as possible that use a circle, or other graphic forms. Also, Wallach and Kogan (1965) based their measures of creativity, the Wallach and Kogan Creativity Test (WKCT), on divergent thinking. In more recent psychometric tools, such as the Evaluation of Potential Creativity (EPoC), Lubart, Barbot, and Besançon (2011) include divergent-exploratory thinking tasks as well as converge-integrative tasks. For example, in the artistic-graphic domain, children and adolescents are solicited, for divergence, to create many drawings from a simple stimulus form, and to create one elaborated drawing using several forms in the convergent task. This is clearly compatible with numerous creative process models, including Guilford's work.

In the second line of work, people in applied fields, notably in business, saw the relevance of divergent thinking, convergent thinking and evaluation to help boost creativity. This was the case of Alex Osborn, an advertising executive who had an agency (BBDO). He noticed that his "creatives" who were in charge of generating new ideas for advertising campaigns were not necessarily as creative as they could be. Osborn (1953) observed that they might improve their creative process by structuring it; in particular, Osborn proposed to enhance the divergent thinking phase, and defer evaluation of ideas until later in the process. These two ideas formed part of the rules of "brainstorming". Later, authors such as Sidney Parnes, Donald Treffinger, Scott Isaksen and Gerard Puccio, with many others expanded on these initial ideas to develop the "Creative Problem Solving" (CPS) process. CPS structures the creative process, distinguishing divergent, evaluative and convergent thinking, in three main phases (exploring the challenge, generating ideas, and planning for action) (see Isaksen & Treffinger, 1985; Parnes, 1967; Puccio & Cabra, 2009; Treffinger, 1995).

Act 4

From the 1980s, a trend grew to divide complex cognitive processes into constituent components. This approach was successfully used in research on standard intellectual tasks (see Sternberg's (1985) seminal componential analysis work). The study of the creative process focused on the multiple sub-processes that enter into creative thinking. Problem definition, selective encoding of information, mental representation, metaphorical and analogical processes, associative thinking, selective encoding, selective comparison, selective combination are some of the sub-processes that can be identified. Each sub-process itself is composed of more elementary processes. Thus, the creative process is an orchestrated symphony of more specific processes that play together, or in sequence as part of the whole. Concerning convergence, or the synthesis and selective combination of information, bisociation, Janusian thinking, homospatial thinking, articulation, analogy and metaphor, remote association, emotional resonance, and feature mapping are relevant sub-processes (see, for example, Koestler, 1964; Lubart & Getz, 1997; Mednick, 1962; Rothenberg, 1979, 1986, 1996, 2011; Weisberg, 1993). In terms of analytic-evaluative processes, there is work within artistic, literary, and organizational settings (see Runco, 1995). Other sub-processes hypothesized to play a role in creativity have also been investigated, such as perception and information encoding, using decision heuristics, memory and forgetting.

The natural question which arises in a componential approach is how the various sub-processes fit together. For Guilford, the model involves an initial stage of filtering (attention aroused and directed), a stage of cognition (the problem is sensed and structured), a stage of production (ideas are generated with divergent and convergent thinking involved), followed by another cycle that can continue until the task is completed (Guilford, 1967; Merrifield, Guilford, Christensen, & Frick, 1962; Michael, 1999). The order of the stages may vary somewhat and a subset of stages may be repeated in a mini-cycle. The process ends when the work is evaluated as finished, or as unsolvable or finally outdated. (Guilford, 1979). Of course, the extent to which a process is ever really finished is a subject of debate, as it may enter into a new incubation mode and later come back in active work.

Mumford, Mobley, Uhlman, Reiter-Palmon, and Doares (1991) proposed a set of core processes, which enter a dynamic cycle. Problem construction, information encoding (and retrieval), category search (specifying relevant information schemas), specification of best fitting categories, combination and reorganization of category information to find new solutions, idea evaluation, implementation of ideas, and monitoring are core processes. Each can be decomposed into a more specific sub-process model. Mumford and his colleagues examined several of the proposed processes in a series of studies and showed that beyond general ability measures (e.g., grade point average, Scholastic Assessment Test scores, divergent thinking), different processes mentioned here (problem construction, information encoding, category selection, and category combination) explained variance in creative performance on problem-solving tasks concerning advertising and managerial or public policy issues (Mumford, Supinski, Baughman, Costanza, & Threlfall, 1997).

In a somewhat different approach, Finke, Ward, and Smith (1992) proposed the geneplore model; here there is a cyclical movement between generative and exploratory processes. Pre-inventive structures result from knowledge retrieval, idea association, synthesis, transformation, and analogical transfer. Then, exploratory processes come into play to examine, and elaborate these pre-inventive structures. This is similar to several other "models" that postulate idea generation and idea evaluation in dynamic sequences (see Runco & Chand, 1995). In the psychodynamic approach, primary and secondary processes come into play (Kris, 1952; Kubie, 1958; Suler, 1980). The primary process operates on unstructured, illogical, subjective thoughts and yields ideational material that is then shaped by the reality-based, controlled, evaluative secondary process. In the BVSR model, previously mentioned, there is idea generation (through random combinatory processes) and then active evaluation and retention (Campbell, 1960; Simonton, 2011). In the "creative problem-solving" (CPS) tradition, Basadur's (1995) work mentions specifically ideation–evaluation cycles that vary in their frequency according to the nature of the problem to be solved and the point in problem-solving (e.g., at the beginning vs. in a final implementation phase).

In terms of recent trends, there have been many studies of brain activity during cognitive tasks that focus on divergent thinking, remote association, mental flexibility, or insight problem-solving. These studies have extended the knowledge of sub-processes involved in creativity, and specified in some cases the brain structures or networks that are solicited in each sub-process (see Vartanian, Bristol, & Kaufman, 2013). In applied settings, a large number of creativity techniques, and some studies of their efficiency, have developed. These techniques tend to focus on specific cognitive processes, such as analogical thinking, mental flexibility and heuristics (TRIZ, ASIT, lateral thinking techniques, Atshuller & Seredinski, 2004; De Bono, 2009) or conceptual knowledge reorganization (C-K techniques, Hatchuel & Weil, 2009). The development of creative process models continues and the dynamic aspects of such models have been recently receiving more attention (Botella & Lubart, in press).

The Creative Process—Types of Research

Research on the creative process has been mainly descriptive in nature. Studies have reported observations of how creative activity unfolds. These observations have sometimes been made by those engaged in the creative act, through self-report by the creators themselves, or by external observers. For example, Israeli (1962) reported on his own creative painting activity in one early published study. Getzels and Csikszentmihalyi (1976) observed art students who made still-life drawings, counting the time each participant spent on various artistic activities, and documenting by photographs how the drawings evolved. They found that process-related variables, such as the initial time spent exploring the still-life objects, called problem finding, were related to the creativity of the artists' work on the drawing task and the artists' later careers.

Some researchers have interviewed creative people who described their work process, and other scholars examined traces of creative activity, such as manuscripts of authors which show revisions of the text as it was composed. An early example of these methodologies can be found in Alfred Binet's studies of dramatic authors, who described their creative writing process in structured interviews, and provided in some cases examples of their handwritten manuscripts which Binet analyzed in terms of evidence for revisions (Binet & Passy, 1895). Gruber and Barrett's (1974) well-known studies of Charles Darwin's notebooks from his voyage on the Beagle allowed Darwin's creative thinking underlying the theory of the Evolution of the Species to be traced.

Some studies have, however, taken a different approach to the creative process. Using an experimental paradigm, people are instructed to work under various conditions, to engage in the creative process in a certain way, following certain steps, and the quantity, quality or the content of the resulting productions is examined. In general, if the process variables make a difference, the resulting productions should be more or less creative. An example is research on the effects of prompting authors to evaluate their ideas at various moments during the writing process, and examining its' effect on the creativity of the work produced (Lubart, 2009). In a set of studies, university students were asked to create short stories based on a title or set of characters. Some students were prompted to evaluate their work in progress early in their writing process, others were prompted toward the end of the process, and some students were prompted on a regular schedule throughout the work period. In addition, there were control conditions with students who received no prompts, or were prompted to evaluate on a schedule but the evaluation was not related to the story composition task. Among the results, the study showed that the early evaluation condition led to more original stories, which may be due to students avoiding to quickly settle on their first, often common story ideas.

Each of these methodologies, be they observational, experimental, qualitative or quantitative, can offer insights on the creative process. It is also worth noting some studies look at the creative process in natural settings, such as professional work environments, whereas others have participants engage in creative thinking on demand, in somewhat artificial laboratory settings.

The Nature of the Creative Process: Putting It All Together

Empirical studies that compare, in detail, how people engage in creative work as part of their daily jobs are surprisingly rare. Based on the concepts and historical overview presented in this brief introduction to the creative process, a large-scale research project was undertaken to describe and model the creative process in several domains of professional creative activity. Several of the authors who contributed to the current edited volume joined together to examine the creative process in visual arts, writing, science and engineering, design and musical composition, which are representative of a range of real-life creative work domains. It was hypothesized that these domains may have descriptions of the creative process which overlap, but also show some unique facets. Finally, it may be possible to enhance creativity by training people to engage in certain process steps. Thus, there are important educational implications from research, which can help students or professionals to enhance creativity through process training.

Multiple Methodologies: Interviews, Notebooks, Qualitative and Quantitative Analyses

In a first set of studies, professionals recognized for their creative work (prizes, accomplishments, etc.) were queried using a structured interview technique. They described their creative process. These interviews were analyzed qualitatively and quantitatively for use of certain terms related to the creative process. Models of discourse were conceived and the five professional domains were compared.

In a second set of studies, advanced students in the same professional fields engaged in creative projects that lasted several sessions (and covered several weeks). The compositions were sculptures, film scripts, engineering solutions, musical compositions, and product or poster designs. As participants worked, they completed a structured observation notebook that traced their activities and thinking. These were analyzed qualitatively and allowed specific sequences of actions to be detected. In particular, students whose works were judged as highly creative were contrasted, in terms of their creative process with other participants who were not recognized as especially creative.

Finally, in a third set of the studies, students were identified based on their creative abilities and provided training to enhance their creative process. This work used different pedagogical treatments; students were pretested for specific types of skills that had been identified as particularly involved in the creative process. Those who had low scores on these skills were then provided training to enhance the use of the skills during the creative work. Thus, an experimental design was used to examine if instruction in process-related skills would lead to better performance compared to control group participants who received unrelated training.

The main results of the first phase focused on interview data. Specific terms used to describe the creative process in each field, the frequency of these terms and the related words that represent the semantic field were identified. Then, similarities and differences across creative domains were examined qualitatively. In the end, an action-based theory was used to model the creative process across fields, showing the complex set of similarities and differences (see Glăveanu et al., 2013) The second phase yielded a data set in which the frequency of process activities, the sequences of activities, and the related cognitive and emotional characteristics associated with these activities led to a description of the creative process for each domain. Then, comparisons were made between participants within each domain, to see if there were differences between the creative process sequences for high and low performers, based on judges' ratings of the participants' output creations. Analyses revealed differences in each domain, with process sequences that were specific to the C+ (more creative) and C− (less creative) groups. Domain specific effects were found when examining the creative process. The third phase involved process training. The results showed that specific process components could be identified, measured and trained. Participants showed changes in their work patterns before and after training and some improvement in their creative output.

This research project served as a starting point for the current volume. The structure of the volume presents work on the creative process organized by field of endeavor. The disciplines of visual arts, literary composition, science and engineering, design and music are represented. The current volume examines the creative process in each domain, by including findings from the literature, and new insights gained through multiple methodological approaches.

Bibliography

Altshuller, G., & Seredinski, A. (2004). *40 Principes d'innovation TRIZ pour toutes applications*. Paris: Seredinski (Avraam). EAN.

Basadur, M. (1995). Optimal ideation-evaluation ratios. *Creativity Research Journal, 8*(1), 63–75.

Beaty, R. E., & Silvia, P. J. (2012). Why do ideas get more creative across time? An executive interpretation of the serial order effect in divergent thinking tasks. *Psychology of Aesthetics, Creativity, and the Arts, 6*, 309–319.

Berland, E. (2013). *Barriers to creativity in education: Educators and parents grade the system*. San Jose, CA: Adobe Systems.

Berman, S., & Korsten, P. (2010). *Capitalizing on complexity: Insights from the global chief executive officer study*. Somers, NY: IBM.

Binet, A., & Passy, J. (1895). Notes psychologiques sur les auteurs dramatiques. *Année Psychologique, 1*, 60–118.

Botella, M., & Lubart, T. (in press). From dynamic processes to a dynamic creative process. In R. Beghetto & G. Corazza (Eds.), *Dynamic perspectives on creativity: New directions for theory, research, and practice in education*. New York: Springer.

Campbell, D. T. (1960). Blind variation and selective retention in creative thought as in other knowledge processes. *Psychological Review, 67*, 380–400.

De Bono, E. (2009). *Lateral thinking: A textbook of creativity*. New York: Penguin.

Eindhoven, J. E., & Vinacke, W. E. (1952). Creative processes in painting. *Journal of General Psychology, 47*, 165–179.

Finke, R. A., Ward, T. B., & Smith, S. M. (1992). *Creative cognition: Theory, research, and applications*. Cambridge, MA: MIT Press.

Getzels, J., & Csikszentmihalyi, M. (1976). *The creative vision: A longitudinal study of problem finding in art*. New York: Wiley Interscience.

Ghiselin, B. (Ed.). (1985). *The creative process: A symposium*. Berkeley: University of California Press (Original work published 1952).

Glăveanu, V. P., Lubart, T., Bonnardel, N., Botella, M., De Biasi, P.-M., De Sainte Catherine, M., ... Zenasni, F. (2013). Creativity as action: Findings from five creative domains. *Frontiers in Educational Psychology, 4*, 1–14. https://doi.org/10.3389/fpsyg.2013.00176.

Gruber, H. E., & Barrett, P. H. (1974). *Darwin on man: A psychological study of scientific creativity*. E. P. Dutton: New York.

Guilford, J. P. (1950). Creativity. *American Psychologist, 5*, 444–454.

Guilford, J. P. (1967). *The nature of human intelligence*. New York: McGraw-Hill.

Guilford, J. P. (1979). Some incubated thoughts on incubation. *Journal of Creative Behavior, 13*, 1–8.

Hatchuel, A., & Weil, B. (2009). C-K design theory: An advanced formulation. *Research in Engineering Design, 19*(4), 181–192.

Hitt, W. D. (1965). Toward a two-factor theory of creativity. *Psychological Record, 15*, 127–132.

Isaksen, S. G., & Treffinger, D. J. (1985). *Creative problem solving: The basic course*. Buffalo, NY: Bearly Limited.

Israeli, N. (1962). Creative processes in painting. *Journal of General Psychology, 67*, 251–263.

Koestler, A. (1964). *The act of creation*. New York: Macmillan.

Kris, E. (1952). *Psychoanalytic exploration in art*. New York: International Universities Press.

Kubie, L. S. (1958). *Neurotic distortion of the creative process*. Lawrence: University of Kansas Press.

Lubart, T. I. (2000). Models of the creative process: Past, present and future. *Creativity Research Journal, 13*(3–4), 295–308.

Lubart, T. (2009). In search of the writer's creative process. In S. B. Kaufman & J. C. Kaufman (Eds.), *The psychology of creative writing* (pp. 149–165). New York: Cambridge University Press.

Lubart, T. (2017). The 7 C's of creativity. *Journal of Creative Behavior, 51*(4), 293–296.

Lubart, T. I., & Getz, I. (1997). Emotion, metaphor and the creative process. *Creativity Research Journal, 10,* 285–301.

Lubart, T. I., Besançon, M., & Barbot, B. (2011). *Evaluation du Potentiel Créatif (EPoC). (Test psychologique et Manuel)* [Evaluation of creative potential: Test and manual]. Paris: Editions Hogrefe France.

Mednick, S. A. (1962). The associative basis of the creative process. *Psychological Review, 69,* 220–232.

Merrifield, P. R., Guilford, J. P., Christensen, P. R., & Frick, J. W. (1962). The role of intellectual factors in problem solving. *Psychological Monographs, 76*(10), 1–21.

Michael, W. B. (1999). Guilford's view. In M. A. Runco & S. R. Pritzker (Eds.), *Encyclopedia of creativity* (Vol. 1, pp. 785–797). San Diego, CA: Academic.

Mumford, M. D., Mobley, M. I., Uhlman, C. E., Reiter-Palmon, R., & Doares, L. M. (1991). Process analytic models of creative capacities. *Creativity Research Journal, 4,* 91–122.

Mumford, M. D., Supinski, E. P., Baughman, W. A., Costanza, D. P., & Threlfall, V. (1997). Process-based measures of creative problem-solving skills: V. Overall prediction. *Creativity Research Journal, 10,* 73–85.

Osborn, A. F. (1953). *Applied imagination: Principles and procedures of creative problem-solving.* New York: Scribner's.

Parnes, S. J. (1967). *Creative behavior guidebook.* NewYork: Scribners.

Patrick, C. (1935). Creative thought in poets. *Archives of Psychology, 178,* 1–74.

Patrick, C. (1937). Creative thought in artists. *Journal of Psychology, 4,* 35–73.

Patrick, C. (1938). Scientific thought. *Journal of Psychology, 5,* 55–83.

Poincaré, H. (1985). Mathematical creation. In B. Ghiselin (Ed.), *The creative process: A symposium* (pp. 22–31). Berkeley: University of California Press (Original work published 1908).

Puccio, G., & Cabra, J. (2009). Creative problem solving: Past, present and future. In T. Rickards, M. A. Runco, & S. Moger (Eds.), *The Routledge companion to creativity* (pp. 327–337). London and New York: Routledge and Taylor & Francis.

Reiter-Palmon, R., Mumford, M. D., O'Connor Boes, J., & Runco, M. A. (1997). Problem construction and creativity: The role of ability, cue consistency, and active processing. *Creativity Research Journal, 10,* 9–23.

Rothenberg, A. (1979). *The emerging goddess: The creative process in art, science and other fields.* Chicago: University of Chicago Press.

Rothenberg, A. (1986). Artistic creation as stimulated by super-imposed versus combined-composite visual images. *Journal of Personality and Social Psychology, 50,* 370–381.

Rothenberg, A. (1996). The janusian process in scientific creativity. *Creativity Research Journal, 9,* 207–231.

Rothenberg, A. (2011). Janusian, homospatial and sepconic articulation processes. In M. A. Runco & S. R. Pritzker (Eds.), *Encyclopedia of creativity* (2nd ed., Vol. 2, pp. 1–9). New York: Elsevier.

Runco, M. A. (Ed.). (1995). *Critical creative processes.* Cresskill, NJ: Hampton.

Runco, M. A., & Chand, I. (1995). Cognition and creativity. *Educational Psychology Review, 7,* 243–267.

Runco, M. A., & Jaeger, G. J. (2012). The standard definition of creativity. *Creativity Research Journal, 24*(1), 92–96.

Simonton, D. K. (2011). Creativity and discovery as blind variation: Campbell's (1960) BVSR model after the half-century mark. *Review of General Psychology, 15*(2), 158–174.

Stein, M. I. (1974). *Stimulating creativity: Individual procedures.* New York: Academic.

Sternberg, R. J. (1985). *Beyond IQ: A triarchic theory of intelligence.* Cambridge: Cambridge University Press.

Suler, J. R. (1980). Primary process thinking and creativity. *Psychological Bulletin, 88,* 144–165.

Torrance, E. P. (1988). The nature of creativity as manifest in its testing. In R. J. Sternberg (Ed.), *The nature of creativity* (pp. 43–75). New York: Cambridge University Press.

Treffinger, D. J. (1995). Creative problem solving: Overview and educational implications. *Educational Psychology Review, 7,* 301–312.

Vartanian, O., Bristol, A. S., & Kaufman, J. C. (Ed.). (2013). *Neuroscience of creativity.* Cambridge, MA: MIT Press.

Wallach, M., & Kogan, N. (1965). *Modes of thinking in young children.* New York: Holt Rinehart & Winston.

Wallas, G. (1926). *The art of thought.* New York: Harcourt Brace.

Weisberg, R. W. (1993). *Creativity: Beyond the myth of genius.* New York: Freeman.

2

How Do They Do It? The Importance of Being ... eArNe(ARTI)ST

Ivan Toulouse

The creative process could be represented as a tree. At each point, the artist is facing a choice among several branches, then, after a while, facing a new indetermination and new choice, and again and again… On the other side, the curator who organizes a retrospective has an interest in Picasso's drawings during his childhood, for instance, because the painter was well known afterwards. Otherwise nobody would have ever paid any attention to them. Viewing the work from the end is like Ariadne's thread. It can lead Theseus back to the entrance after having killed the Minotaur, and then escape the Labyrinth. There is no longer any decision to be made between two or three possible routes. You just have to follow down the main way, back to the trunk, and back to the beginning. That does disconnect all the tensions of what was an undetermined process, and that makes it look as simple and obvious as a

---

A tribute to Oscar Wilde.

I. Toulouse (✉)
Rennes 2 University, Rennes, France

© The Author(s) 2018   **19**
T. Lubart (ed.), *The Creative Process*, Palgrave Studies in Creativity and Culture,
https://doi.org/10.1057/978-1-137-50563-7_2

rectilinear route, as soon as the chronological order has been restored. Thus the visitor of the Picasso Museum, in Paris, finds it totally logical that the artist had been passing through these different periods (fauve, rose, blue) before entering cubism... and nobody even worries about the fact that around 1917 he still paints cubist canvasses at the same time as a neoclassical portrait of his wife Olga sitting in an armchair.

It is like an outward journey compared to the return: if I want to reach a remote little village in the country, for instance, I will probably drive the wrong way and "recalculate" my route quite often, if not get lost twice or more, whereas, for the return, I just have to follow the signposts "Paris" from the start and I shall be home soon without any trouble. The artist is on the way "out" and the onlooker (or the curator) on the way back.

To provide a modelization of "the maze of a creative search", as the artist experiences it, Anton Ehrenzweig proposes that there are multiple paths leading from the starting point. Each one opens upon a nodal point that again has many paths available. Some paths lead to dead-ends, or open ends that do not seem to end.

> The creative thinker has to advance on a broad front keeping open many options. He must gain a comprehensive view of the entire structure of the way ahead without being able to focus on any single possibility. (Ehrenzweig, 1991, p. 70)

The creative process would require considering a massive amount of possibilities from "nodal points" which mark out the paths. And each one of these choices has a crucial importance as on the later development of the work. And yet the artist has no aerial overview of this network. Neither, is there a map of it. As Ehrenzweig says:

> If we could map out the entire way ahead, no further search would be needed. As it is, the creative thinker has to make a decision about his route without having the full information needed for his choice. This dilemma belongs to the essence of creativity. (Ehrenzweig, 1991, p. 70)

In other words, the linear retrospective and explanatory view has very little to do with what the artist's creative process is in reality.

Various Moments of Creation

Involved in a creative research, the artist, according to Ehrenzweig, faces first some fragmentation. What happens does not fit with what he/she wanted, leading to a feeling of insatisfaction. This is an unavoidable first stage of creation and this "fragmentation" is a reflection of the fragmentation of the artist's own personality. Things do not go as wanted. One could say it is the collapse of secondary process. It is the *schizoid* phase. It could be illustrated by the romantic image of the painter "fighting" with the canvas!

A second step is "unconscious scanning." This is the *maniac* phase. Without however abolishing disruption, although fragmentation at the surface does disappear, a global unconscious substructure appears that links the fragmented elements in a syncretic way and that reveals itself as the matrix of the work. Thus "hidden order" emerges. For Ehrenzweig, this is not necessarily a state of pathological regression, but the result of an extreme *dedifferentiation* that characterizes a process of creation. This phase could match the mythology of inspiration.

The third moment is a *re-introjection* of a part of the unconscious substructure of the work in the inner self of the artist at a higher mental level. One could even say that the work imposes itself upon the artist, as far as he or she cannot justify it, because it still does not correspond to what was expected at first. But it has to be coped with in the hope of a future integration. That is the reason why this further stage is a *depressive* phase. It comes along with an anxiety that is connected with *melancholy* which is consubstantial with the idea of *genius*. Little by little this new coherence found by the artist will be consciously adopted through the secondary process of reflection and elaboration.

There is a fourth moment which Ehrenzweig does not talk about and which is for the artist the moment of exhibition. This would correspond to verification for the mathematician. It is the phase when the work is released. It is not only the matter of the reception by the spectator nor the reaction of the public which impacts the artist's behavior which would be another point. Just regarding the process of creation, by exposing his or her work, the artist exposes himself or herself to criticism, to negation. There is danger. But by getting free of him—or of her, as a newborn child becomes autonomous by parting from his or her mother—the work occurs to the artist in a different and distanced way. In this respect, the studio space and the exhibition space are fundamentally different. The sight of one's own exposed work gives the artist the possibility to validate it. People often worry about the fact it would be difficult for an artist to sell a piece and to part with it even if it brings money. Actually it is rather a liberating satisfaction to see that the work has been recognized and adopted by somebody else. It is the proof it works on its own. And the lack that its departure causes leads the artist to get back into the saddle in order to replace it by something new. The capacity of overcoming such a moment is, I believe, the confirmation of an artistic posture.

It is necessary to insist on the fact that these various moments are not really successive stages but an articulation between various states, various attitudes which alternate in the progress of the process but can also overlap or come together. They can be situated in different time-scales. Thus, for example, the fourth moment of validation can happen as a retrospective in a museum at the end of a period of several years of work, but it can consist, after a couple of weeks of a search, in a studio "*accrochage*," as seen in photographs of Picasso with all the piled up paintings hanging above one another. It can be, at any moment, after a few minutes of action, the three steps backwards from the emerging creation, just to see how it works.

Unconscious Scanning

To begin, it was as a psychologist-clinician that Ehrenzweig inquired about art as an object of study, to attempt to reconsider the concept of "primary process." The appearance of disorder masks a "hidden order of art" which he tries to highlight, in the same way that Freud had showed, in his *Traumdeutung*, how the latent content of a dream can be

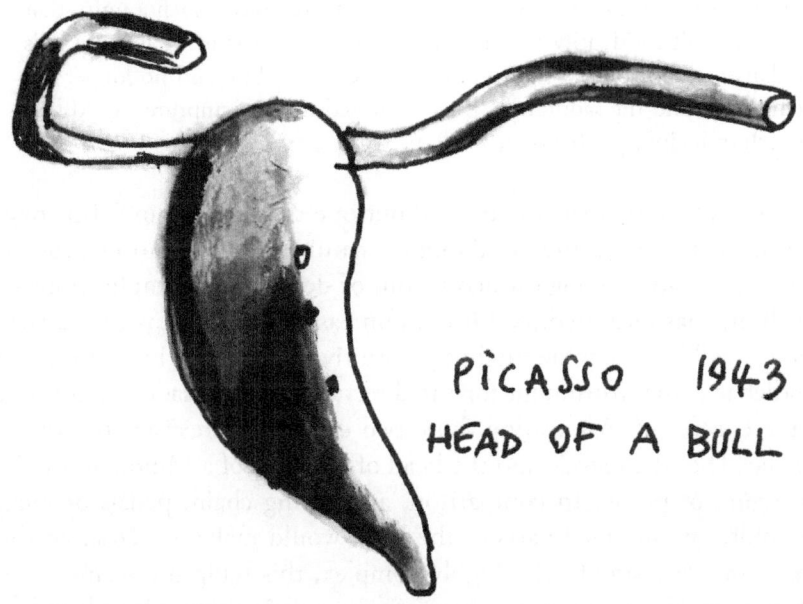

PICASSO   1943
HEAD OF A BULL

Fig. 2.1 Illustration of Picasso's 1943 "Head of a Bull" (illustration by Ivan Toulouse)

interpreted from its manifest content, which apparently is often absurd, by taking account of the phenomena of displacement and condensation in particular.

Not being able to consider, one after another, all the much too numerous possibilities, the artist can only rely on a kind of vague intuition which is going to guide him or her to an interesting solution—and that is presumably what characterizes an artist. This is what Ehrenzweig calls "unconscious scanning." This could be illustrated by this quote from Picasso, commenting to photographer Brassaï his famous *Tête de taureau (Head of a Bull)*, made merely by assembling a bike saddle and handlebars, melted out of bronze (Fig. 2.1):

The idea of this *Head of a Bull* occurred to me without me thinking it out... I did nothing but weld them together... What is marvellous in bronze is that it can give to the most heteroclite objects such a unity that it is difficult to identify the elements from which it is made up. But it is a danger too; if one would only see the head of a bull and no longer the handlebar and the saddle which have formed it, this sculpture would lose much of its interest. (Brassaï, 1969, p. 76)

This was an "idea" that occurs "without one thinking it out." This oxymoron characterizes the "unthought" nature of the creation process. It is not a matter of logical deduction, of decision, the implications of which one has measured. No! It is an unreasoned dash. However, a later analysis will be able to highlight the intuitive logic of the lucky find: the handlebar is the instrument for the direction, and the saddle is the seat of the command. Associating those two elements is relevant to express the "head" of the bicycle and the head of a bull or of a Minotaur, mythical figure of power. In comparison, assembling chain, pedals or rack, which are the subdued parts of the bike, would make no sense. At the same time very simple and highly complex, this sculpture seems to be a paradigm of a piece of art, because the artistic gesture is reduced to its very least: no physical commitment, no material transformation, no use of any techniques... and yet the criteria of creation are fulfilled. Art is here merely typified as creative energy. According to René Passeron (1993), the three criteria of creation are: (1) the production is not mainly directed toward a functional goal; (2) it has a symbolic range as if it were a pseudo-person; and (3) and its' author is emotionally involved in it.

Let us note, by the way, as Picasso says himself, that the disparity of materials is reunified here by bronze: each element is decontextualized from a former structure—the bicycle, where it has its function from which is only preserved an essential meaning—then recontextualized into a new organism—a sculpture—which takes its autonomy. The use of bronze is here a paradigm of how such an *assemblage* of various objects or materials into a piece of art typifies the passage from metaphor to metamorphosis. The choice operated by the artist is not conscious but obeys a kind of "elective affinities." Generally speaking, creation is not led by an open-air intention but by an underground determination. Picasso said:

> If you give certain things a meaning in my paintings, it might be very true, but I never had thought of it. The ideas and conclusions you draw, I came to them too, but instinctively, unconsciously. I paint things for what they are. It is in my subconscious mind. (in Picasso, 1998, p. 51)

In contrast to conscious focusing, unconscious scanning is the ability to move in a kind of "disrupted" or dispersed consciousness. Of course, it is not a permanent state. There is something like an alternation between moments of focusing and breaking up, of converging and diverging. But the artistic work presumes this ability to switch, to disconnect, to empty oneself, to let go, which has something to do with the myth of the visionary or prophetic artist. And, as seen above for science as well, it is not the concern of artists only.

> The scientist has to face the fragmentation of physical facts with courage. He has to scan a multitude of possible links that could make sense out of apparent chaos. I would maintain that he needs the more dispersed (undifferentiated) structure of low-level vision in order to project the missing order into reality. [...]. The artist, too, has to face chaos in his work before unconscious scanning brings about the integration of his work as well as of his own personality. (Ehrenzweig, 1991, p. 39)

Quite often random appearance in some ways of processing is not only misleading but the creator himself is conscious that it has nothing to do with chance; consider the drippings of Jackson Pollock, for instance. Having emigrated during the war, French surrealist painter André Masson introduced young New Yorkean artists to the "automatic writing" of his poet friends that he had transposed to drawing and painting, and Pollock brought it up to a larger scale. Being asked if it was not more difficult to control his dripping than a brush, Pollock, seeking his words, answered:

With experience it seems to be possible to control the flow of paint, to a great extent, and I don't use – I don't use the accident – 'cause I deny the accident… it's quite different from working, say, from a still life where you set up objects and work directly from them. I do have a general notion of what I'm about and what the results will be. I approach painting in the same sense as one approaches drawing, that is, it's direct. (Harrison & Wood, 1992)

Concentration, Emptiness, and Fullness

This idea of floating attention in the mind of the artist and this *subtle* (from Latin *sub-tela*, under the cloth) conception of human spirit might seem far away from our occidental way of thinking but is a truism in oriental wisdom. Concerning yoga, transcendental meditation, taoïsm, or zen philosophy, emptiness is a familiar notion and concrete reality. As an example, for the Chinese, emptiness or vacuum, *wu*, is not, as it is for us, void or nothingness, that is to say absence but, on the contrary, it is a kind of absolute presence that makes fullness possible. It is the origin of what Ehrenzweig points out as "indifferentiation." Without entering the details of this question that has been developed in its whole complexity by François Cheng (1991), let us just recall this aphorism of Lao-Tzu saying that if a vase is made of clay, it is its emptiness that allows to fill it.

Together with emptiness, there is an attitude of quietness, *hsü*, to which every being should aim and which could be related to the state of *ataraxy* advocated by Epicureans, and that could be defined as a state of no need, of indifference, beyond desire. It could be compared to the *virtue* of Stoïc philosophers like Seneca, for instance, as well. Those two philosophical attitudes have often been opposed to each other whereas they seem to be very similar. It might even have something to do with the "*beauté d'indifférence*" (beauty of indifference) that Marcel Duchamp sought in his *ready-mades*.

The type of concentration necessary for an artist is not sustained attention, but a kind of loose control. Many artists confirm this. Let us take briefly three examples in the past century: Klee, Matisse, Dubuffet.

In *The Thinking Eye*, Paul Klee offers some guidance to his students at the Bauhaus for their first lesson:

> As a negative pole, chaos is not, properly speaking, intrinsic chaos but it represents a notion in a precise position, defined in relation with the notion of cosmos. Chaos, at its proper meaning, never could take place on the scales; it will stay eternally imponderable and unmeasurable. It can be nothing or be "something" at the state of half-sleep, death or birth, according to the data: intention or lack of intention, will or negation of will. The symbol of this "non-notion" is the point. It is not, to tell true, a real point but the mathematical point. This something-nothing, or even this materialized nothingness, is an unconceivable concept, characterized by absence of contrast. If it is given a perceivable significance (that is to say if finality it is introduced inside chaos), one gets the concept of grey. (Klee, 1973, pp. 3–4)

Striking is the correspondence between this explicitation of the visual metaphysics as proposed by Klee's paintings and this description of a kind of prerational thought. The emphatic tone in which it is stated could seem esoteric and doctrinal, even irritating, but it is shared by most of the "avant-garde" artists (as Malevich, Kandinsky, or Mondrian). Actually there are similarities with the mental visualization on which Tibetan *hatha-yoga* exercises are based, for example, in which concentration through blankness is practiced under the name of *ékâgratâ*: the spirit intensely controls *indriyas*, which are the subtle principles of the efficiency of our five senses. By the way, the great specialist of oriental religions, Jean Varenne, related this anecdote concerning Henri Poincaré, known for his writings on the mathematical creative process:

> Well known is this story of a mathematician who, lost in his thoughts, found himself, while walking in Paris, facing a horse-carriage whose back, covered by a stretched tarpaulin, appeared to him as a perfect blackboard. Drawing a piece of chalk out of his pocket, he started to write down the data of the problem he was thinking about, unaware of this unusual situation; and when the carriage started, the scientist was seen running behind it to pursue the procedure of his demonstration. This is what yogins name *ékâgratâ*, which is not distraction of a rather

non-down-to-earth intellectual but, on the contrary, attentionate work of the mind focusing its power on one and only one object. (Varenne, 1989, p. 121)

Matisse described his own process in the following way:

> I understood that the composing mind must keep a sort of virginity about the chosen elements and reject what comes to him by reasoning. [...] After having blanked my brain, emptied it from preconceived idea, I would be drawing this preliminary indication, with a hand that was only guided by my unconscious sensations, stemming from the model. I would avoid carefully introducing in this representation any intentional remark or rectifying a material mistake. The almost unconscious transcription of the model significance is the initial act of any work of art and particularly a portrait. (Matisse, 1972, p. 178)

As described by Jean Dubuffet, an overly directional approach to reality will only focus on a detail. He recommends a global input. It is the paradox of the tree that hides the forest.

> Attention kills what it touches. It is an error to believe that to look at things attentively you get to know them better. Because the eye spins like the silkworm so that, in an instant, it wraps itself inside an opaque cocoon which deprives you of any sight. That is why painters who stare at their model don't catch anything of it at all. (Dubuffet, 1967, p. 61)

In Dubuffet there is however a supplementary paradox: there is no "oriental" wisdom like in Klee or Matisse, but void is obtained through saturation of a quite "occidental" overflowing, just like the profusion of our consumer society. Thus, he seems to have used as a meditation chamber his *cabinet logologique*, that he set up in *Villa Falbala*, which he had designed in Périgny-sur-Yerres. There the walls are covered with characteristic drawings of his "Hourloupe" period (in the sixties). In these drawings, drawn in black and sometimes streaked with blue or red hatchings, overlap and combine with one another various objects and uncountable figures, the proliferation of which makes them eventually disappear.

Intuition

Marcel Duchamp in 1957, at the venerable age of 70, said:

> To all appearances, the artist acts like a mediumistic being who, from the labyrinth beyond time and space, seeks his way out to a clearing.

> If we give the attributes of a medium to the artist, we must then deny him the state of consciousness on the esthetic plane about what he is doing or why he is doing it. All his decisions in the artistic execution of the work rest with pure intuition and cannot be translated into a self-analysis, spoken or written, or even thought out. (Duchamp, 1994, pp. 188–189)

This could allow us to think that, in his youth, in 1913, this trailblazer of *conceptual* art, did not "think out" his first *ready-made*; it was neither the result of a reasoning nor did it come from a deliberate decision, but perhaps occurred simply as a wonder at the sight of passing-by bicycle wheel spokes. According to Duchamp as for many artists, creation does not result from rational intelligence. It comes from intuition. Some even claim that creation is only possible if it bypasses intelligence.

For Marcel Proust it is only "sensation" not intelligence that can lead to knowledge, to truth. Everybody knows the famous "episode of the *madeleine*," in which the taste of the biscuit activates an involuntary memory. And oddly enough, it is the most intimate sensations (gustatory and tactile almost, auditory, a little) which permit this "resurrection"; never those more controlled, more rational, more "intelligent" processes. And this does not only concern creation in art but any human activity, whatever it is, from driving an automobile to researching in sciences. The mathematician Henri Poincaré wrote:

> Logic is not enough, [...] the science of demonstration is not all of science and [...] intuition must preserve its role as a complement, I was going to say as a counterweight or a counterpoison to logic. [...] if it is useful to the student, and still much more to the creative scientist. (Poincaré, 1905/2003)

This form of prerational intelligence, this kind of sensitive way of thinking is particularly obvious in an artistic experience and there is a strange paradox between the difficulty for the artists to explain why they do something rather than something else, while doing it, sometimes even claiming for the unutterability of their creation, and, on the other hand, the kind of confidence in which they seem to be acting most of the time, as if they knew the reason why. In this vein, the American painter Mark Rothko said:

> Intuition is the height of rationality. Not opposed. Intuition is the opposite of formulation, of dead knowledge. (Rothko, 2007, p. 132)

Afterwards things happen to be obvious and logical and a few decades later the most independent artist or even the most rebel one appears to be the "witness of his or her time," as the very representative example of the society he(she) him(her-) self-thought he(she) was even fighting against. As the word implies, a "*retro*-spective" exhibits the work of an artist as it would be seen in a rearview mirror ("*rétro*-viseur" in French), that is to say from the end toward the beginning. But when the artist was doing his (her) work, he(she) was not viewing it in that way! He(she) was just doing something pragmatically, not knowing really why, and could have chosen another possibility instead of that very one. It is the same with great inventions: it seems obvious that you shall get water when turning your tap! Yet it had to be invented.

An Ethics of Creation

Ehrenzweig refers to an ethics of creation that includes humility, courage, generosity, and love which are not dogmatic virtues one should conform to, but are imperative attitudes to make creation possible.

As the work is partially beyond control, by experiencing creation, the artist is obliged to allow a certain renunciation. This is perceptible from the first phase when the artist sees the project slipping away. It is not reluctant resignation but an active acceptance. French painter Pierre

Soulages suggested that painting required incessantly escaping from the initial project.

This is a difficult paradox to accept. At the same time, the artist has to engage all his or her personality and energy in acting but he or she has to give up checking everything and to put away his or her ego. It is necessary to make a stream of volition efficient while decreasing the power of one's authoritative will! In this respect, Dutch painter Bram Van Velde goes even further when, with an almost taoist accent, he says contradictorily:

> Most live under the reign of the will. The artist is the one who is without a will.[...] Each canvas represents a moment when we could, when we had the strength. (Juliet, 1978)

He has to accept himself/herself as he or she is, with his limits and not as he or she would like to be. Far from a well-anchored prejudice, it is the opposite of a narcissistic attitude. One could not say, for example, that the self-portraits which Rembrandt repeated throughout his life express self-satisfaction. On the contrary they picture his uneasy inquiry about who or what he is. For Ehrenzweig, there is thus in creation an inherent form of humility.

The autonomous life of the work, begins with its material existence. Its materiality imposes rules on which the artist has no influence. The material exists independently of the artist. There is thus a confrontation in which the artist has to make a commitment. For Picasso, something tragic takes place in this confrontation which involves both material and psychic reality:

> For me, to paint a picture, it is to commit a dramatic action during which the reality finds itself torn. This drama gets the upper hand over any other consideration. What matters, it is the drama of the very act, the moment when the universe escapes and faces its own destruction. (Picasso, 1998, pp. 118–119)

The fact has to be accepted that it is only through this confrontation that something will be happening. Whatever the period every piece of art does operate aesthetically, that is to say, produces in the one who contemplates it a sensation (*aisthesis* in Greek). Some conceptual artists claim their concern is outside this confrontation. If it were the case, they would miss the opportunity (*kairos*) of real creation. But it seems to me that, in this extreme artistic attitude—as when one studies the borderline cases of a mathematical function—the confrontation with this sensitivity takes place in several manners: at first in its withdrawal, in an experience of void or silence which is as sensitive as an experience of fullness or noise, although it is more difficult to perceive because it needs more intense concentration. Sensitivity is also necessarily implemented in the means they use, as an electronic bulletin board for Jenny Holzer, for example, or inscriptions on walls for Lawrence Weiner. And finally widening the notion of sensitivity, they make use of the linguistic or institutional codes and work their "plasticity," as a "material" in a "enlarged understanding," as Joseph Beuys reported about art in general. And quite often, artists have to struggle with the medium. For Ehrenzweig, this confrontation appears fundamentally as a frustration.

> The medium, by frustrating the artist's purely conscious intentions, allows him to contact more submerged parts of his own personality and draw them up for conscious contemplation. (Ehrenzweig, 1991, p. 93)

This resistance gives the work in progress the status of something like an *alter ego*. Quoting Adrian Stokes, Ehrenzweig even speaks about the "otherness" of the work of art. It joins moreover the conception of René Passeron who sees in the creative work a "pseudo-person" with whom the artist has necessarily to make "conversation." Jean Dubuffet describes this clearly:

> Art has to arise from the material and from the tool and it has to keep the trace of the tool and the fight of the tool with the material. The man has to speak but the tool also and the material also. (Dubuffet, 1967, p. 57)

Joan Miro expresses very well this idea of a dialogue with the respect due toward an interlocutor:

Obviously; it is necessary to have the highest respect for the material. It is the starting point. It dictates the work. It imposes it. [...] A dialogue exists. It is obvious; a dialogue with the material becomes established. When you make some ceramics, the material of the vase dictates what to do. It imposes its laws. (Charbonnier, 1980)

This thus obliges to accept something else than what the artist expected. Ehrenzweig refers to his experience as a teacher at Goldsmith College of the University of London, where he trained art teachers. He could notice how much trouble they had to accept their own spontaneity and their overflow from stiff programming in their personal artistic works was related with their difficulty to stand the nonetheless fertile unruliness of their young pupils who did not "respect" their instructions.

An immature artist who is hell-bent on exerting full control over his work is incapable of accepting that a work of art contains more than what he had (consciously) put into it. (Ehrenzweig, 1991, p. 145)

Let us come back now to much more pragmatic facts and try to describe the art process in a functional way with the various abilities it requires. To illustrate it, examples from very different periods will be taken that could seem anachronistic, but, despite the different historical contexts, the process may be quite similar. What is sought now is neither a psychological explanation nor an interpretation of the works, but an attempt to report the artist's process and to clarify principles which sometimes sound like concrete instructions of a swimming coach, as the advice that I give to my students in the studio.

Invention, Intention: The Mental Balance

In the course of the process of creation, the artist is confronted with certain recurring difficulties. They appear with variable magnitude and can turn up at any time. The first difficulty to overcome in an artistic experience is to become familiar with the mental gymnastics of the seesaw between exploration and selection, between invention and intention.

When Picasso says : "I don't search, I find," he does not express the conceit of an art hunter boasting he never returns empty-handed from his adventures, as it is often interpreted. Quite the opposite, he testifies to what is undoubtedly at the source of any creation: invention. The word comes from Latin *in-venire*, to come upon, as if by chance. In French, "*inventeur*" is the legal term, as well, that qualifies somebody who happens to find an object in the street, a wreck in the sea. As for the inventors, for the scientists who make discoveries, their inventions often appear to them as a stroke of luck: isn't it the stroke of… an apple which puts Newton on the track to universal gravitation… This is what is often called serendipity.

Artistic inventions show often themselves in this way too. It does not mean that there is no logic, no reason, no idea behind them, but the idea is veiled, in a sense, and the invention just seems to be a chance encounter. Thus, to allow the "hidden order" to be uncovered, to be discovered—another verb to denote invention—it may be necessary, at first, to move away from intention.

At the risk of simplification, we could say that creation begins by exploring a combinatorial space without preconceived ideas. On the opposite, if my action were directly the result of a "design" (from Latin *designare*, to indicate), the result of an "intention" (from Latin, *in-tendere*, to strain), it would point me toward the first—if not the unique—solution which I could have thought of "spontaneously," in other words, by following my habits and prejudices. And among all the uncountable possible solutions, it would certainly not be the best one, or moreover presumably not even a relevant solution at all. If I work in an undecided way, in a flexible state of mind, with "plasticity," if I loosen my approach a bit, I permit a large number of combinations to appear, among which there will necessarily be something more appropriate than the very one toward which I would have straightaway dashed, guided only by the strength of my will. That's why we often say "as luck would have it"! The help of chance constitutes in certain cases, a way of opening the range of possibilities but it does not grant success, by itself. We can compare this with the processes of genetic evolution which obeys the laws of chance. Among all the possible layouts, only the very rare remains which resists bad weather, diseases, and predatory attacks. Nobody questions natural selection since Darwin and, oddly enough, when the subject no longer concerns "procreation" but just "creation," the reasoning seems to no longer apply.

From a very functional point of view, if I were trying to lay out three graphic elements, a title, a text, and an image, for example, the best way would probably not be to stick them after one another upon a big sheet of paper, but to put down the three of them and to move them around to find empirically a display that would seem to me well-balanced or maybe, on the contrary, expressive. And then, but only then, I would fix them. It is what René Passeron calls "formal poietics":

> It could be close to Hjelmslev's glossematics and calculate the combination of the possible institutions from elements given by a system. (Passeron, 1993, p. 438)

In some *collages* by Picasso or Matisse, you can see the needles or thumbtacks implemented to stick the pieces of paper which they are made. It underlines the act of decision which put an end to the open phase of exploration. One of the keys of creation thus seems to be a capacity to act without volition, being guided by intuition through trials and errors. Such a mental attitude is particularly obvious in an artistic work but it is not specific to art.

Choice and Discrimination

If chance can enrich the combinatorial process, it does not mean that all which results from it will be good. Jean Arp, for instance, is said, in 1916, to have thrown in the air small pieces of paper and to have stuck them where they fall. Yet you can see on photographs that the bits did not fall badly, and these collages are powerfully structured. Random sampling did certainly not concern thus the architecture of the collage itself, but only the sequence organization of the scraps of paper. In the same way, it is very difficult to trace spontaneously "any" triangle.

It will be most of the time isosceles if not equilateral or right-angled. Jean Arp doubtlessly wanted the order of the sequence to be arbitrary. But, even by reducing the role of random to that, we are obliged to admit that he made a sorting, and only kept those who seemed suitable to him because they corresponded to this vague idea of a display. This is quite clear if you read carefully what he said on this matter:

> As the arrangement of the plans, their proportions and their colors seemed to depend only on random, I declared that these works were ordered "according to the law of chance," such as in the order of nature, chance being for me only a restricted part of an imperceptible reason for being, of an inaccessible order as a whole. (in Lemoine, 1986, p. 20)

A second type of difficulty lies in not being able to discriminate. It is doubtless that critical discernment intervenes in a selective mode. Intention intervenes—even if it can happen quickly—very little at first as a starting point. This might seem very abstract but it is quite concrete in the material decisions one has to make when painting, for instance.

Thus the posture of creation seems to me to be characterized by this mental seesaw between two attitudes: on the one hand open-mindedness, an availability toward the possible, and, on the other, a closure, a merciless selection. At every moment, the path takes shape. We have already evoked taking three steps backwards of the painter to depolarize the spirit. A small movie shows Antoni Tàpies making a big painting on the ground and getting up almost every 10 seconds to judge his work. Certain tricks are well known, as to look at the picture back to front, or using a mirror which Leonardo Da Vinci recommended

> [...] by painting you have to hold a flat mirror and often look through it at your work; you will see it then inverted and it will seem to you of the hand of another master; so you can better judge its faults than in any other way. (Da Vinci, 1987, p. 260)

The end is then a moment like the others. One does not really know at once if it goes on or if it stops. Pierre Soulages clarifies this moment of discernment in the following way:

When the painting is completed, I turn it to face the wall, and I wait for a long time before looking at it again. I wait moreover longer when I expect it to be disastrous. And when I see it again, in case I have to admit it is true, I remove it from the frame and I destroy it. (Juliet, 1990)

Construction, Destruction, Reconstruction

Let us take the example of Jackson Pollock's paintings. Hans Namuth's famous photographs show his passionate gesture in the studio, projecting paint on the canvas stretched out on the ground. The large format and the horizontal position makes him "plunge" more easily into action. But he knows how to swim! As we saw, he "denies the accident." The work of the unconscious is often mistaken for chance. Ehrenzweig notes also that the dripping and the splashing correspond to an accident:

> There is precious little true accident about dripping and splashing paint. Seen in this way a clever use of accident is as old as art itself. The most skilled techniques of nineteenth-century art knew how to make use of seemingly uncontrollable techniques. The clever water-colourist delights in the untamable spreading of running wet colour. (Ehrenzweig, p. 97)

I would venture to say that some of Pollock's drippings are more successful than others. By questioning this impression, I noticed that those I prefer seem to result from a work in two phases. The initial gesture of the projection expresses its powerful energy but above one can discern covering traces of paint, often in a different color, and sometimes in the shape of wider spots, which, by masking certain parts of the dripping, channel its energy. It works like reframing the many details which, by selecting and by ranking, reconstruct and strengthen the painting. The hypothesis becomes then plausible of a work almost in two phases, clearly recognizable, as the painter himself acknowledges:

> When I am *in* my painting, I am not aware of what I'm doing. It is only after a sort of 'get acquainted' period that I see what I have been about. I have no fears about making changes destroying the image, etc.,

because the painting has a life of its own. I try to let it come through. It is only when I lose contact with the painting that the result is a mess. (Ehrenzweig, p. 97)

We know also that Pollock used to reframe his immense drippings, by cutting them into smaller canvasses. In a lesser proportion, because he made it in the margins, Pierre Bonnard too, used to take back his paintings, by cutting the free canvas he had been working on only tacked on the wall, and by stretching it on a frame just at the end. It is not surprising that for Pollock, afterwards, in 1953, figurative representation even seems to resurface as in *The Deep* where the covering by the white paint redraws it in negative, as a wide gap, and even more obviously still in *Portrait and a Dream* of the same year.

For Picasso, this moment of discernment intervenes in a different manner:

For me, a picture is a sum of destructions. [...] at first, there always is what represents completion for so many others; the masterpiece, the drawing which grabs the world at lightning speed, the docile watercolor in its most subtle delicacies, the gouache and even the oil fixing perfect improvisations; and everything starts with the courage to break this success, to analyze it, that is to say to destroy it in the aim to reach then a designed synthesis, chosen and necessary. (Daix, 1977, p. 21)

A French proverb says: "*Le mieux est l'ennemi du bien*" (*the best is the enemy of the good*). On the contrary, one could express that in art "the good is the enemy of the best." This principle could even be considered a *modus operandi*, particularly perceptible in *The Mystery Picasso (film by Henri Georges Clouzot)* where we see the artist at work. His theatrical side makes him exaggerate certainly. However the systematic character of his method appears clearly. It consists of destroying what has been made to test if it is valid or not. That is also the reason why Picasso speaks of his work process as a "drama." But at the same time, it is a "game." If destruction goes too far, nothing is irreparable: he just plays at it again.

Another example of such a demanding attitude and destruction can be seen in the Sforza castle in Milan with the *Rondanini* Pietà, Michelangelo's last work: it is a marble group representing dead Christ supported by his mother. Only, there is one arm too many, detached from the rest, a right hand of Christ where the muscles are exactly designed and which testifies of a previous state when this sculpture was brilliant and virtuoso, in contrast with the character of incompletion that show the traces of the gradine. All this gives this pietà the divested fervor of a Romanesque sculpture: to tell the truth, you do not know any more if it is Madonna holding Christ in her arms or the opposite that could quite well express the idea of redemption: Christ carrying the suffering of the world on his back, as a father would carry his little child during a walk. In a previous version of the same scene, a few years before, in his *Bandini* Pietà, he represented himself as Nicodeme, a disciple of Jesus who participated in his entombment. This theme is central for Michelangelo, as a subject for meditation, not to speak about his very early Pietà in Rome, when he was twenty five. In the *Bandini* Pietà all his virtuosity appears: he did solve all the difficulties raised by articulating four figures, each one playing a different expressive role. However, already, in a fit of destructive rage, he had been mutilating its Christ whose remnant still shows how he had brought virtuosity to such a peak. A few days before his death—at eighty nine!—he undertook to reshape completely the *Rondanini* such as we see it now, without probably having time to go to the end (see Forcellino, 2006). This Michelangelo's last work is inevitably a kind of testament, a mystical *Less is more,* which testifies to a renunciation of all which could be only vanity, even though it would be the art of the greatest sculptor of his time, if not of all times.

The Fruitful Tension

A third difficulty is linked to the common prejudice about freedom in creation and its romantic corollary: inspiration. Nothing comes from nothing. There is always a sum of determinants that leads us somewhere. However, those determinants do not prevent freedom: we can accept or refuse. An artist is certainly led by an internal necessity, "the inner need," to quote Kandinsky, which cannot be reduced to only satisfying a desire: you cannot say Van Gogh painted for the fun of it! If an artist were only acting to enjoy himself, he would just be a dilettante (*diletto*, in Italien means delight). Kandinsky claimed:

> The inner need is built up of three mystical elements:
>
> 1) Every artist, as a creator, has something in him which calls for expression (this is the element of personality).
>
> 2) Every artist, as child of his age, is impelled to express the spirit of his age (this is the element of style)—dictated by the period and particular country to which the artist belongs (it is doubtful how long the latter distinction will continue to exist).
>
> 3) Every artist, as a servant of art, has to help the cause of art (this is the element of pure artistry, which is constant in all ages and among all nationalities).
>
> A full understanding of the first two elements is necessary for a realization of the third. (Kandinsky, 1912/1969, pp. 109–110)

Most of the time this "internal necessity" will be pushed out by an "external necessity" which is a part of it: a commission, the term of an exhibition, a chance encounter, a proposal to collaborate or—why not?—the constraint of school or university work. There is always a response to contingency without which nothing would ever happen. Otherwise nothing would come out. Instead of complaining about the outer constraints, to be an artist requires to be able to utilize the fruitful tension. It is a necessary but not sufficient condition for a creation to emerge.

Familiar with these principles which are at work in the creation processes, the artist tries then to make them operate. The artist has thus to establish a method that, in many respects, looks like a game. And the game comes along with rules. The artist's method is governed by contradiction. The rule is something closed whereas the game is open.

The Game (and About the Ready-Mades)

A game is a free, an indeterminate activity. In French, we say that two parts which are connected loosely that they "play." The result of the game is never known beforehand. Otherwise there would be no game! The game is a matter of mobility, of plasticity. And the real purpose of the game is a simulation, a delusion. We pretend to aim for the purpose dictated by the rules, but it is not the real purpose. We even think that we play to win but most often we play to play, to make a change, to clear our heads, to have fun (Fig. 2.2).

Let us take an extreme case with the much debated ready-made of Marcel Duchamp, *Fountain* (1917). This urinal in white-glazed earthenware and tipped over in 90 degrees is often given as the counterexample of the work of art in its traditional status as an "auratic"

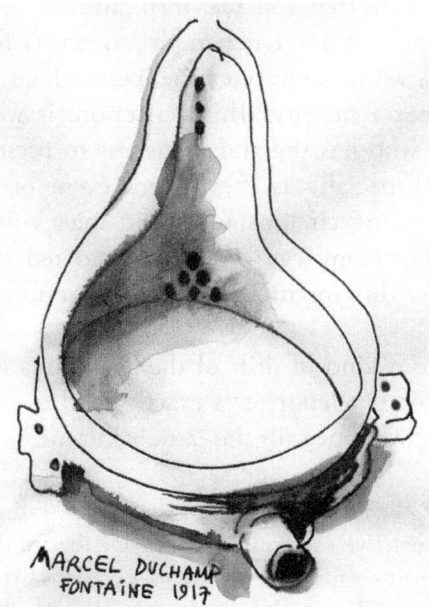

Fig. 2.2 Illustration of Marcel Duchamp's 1917 "Fontaine" (illustration by Ivan Toulouse)

object (Benjamin, 2008). We could very well explain the upgrading of the ready-made object to the rank of work of creation, by resuming René Passeron's criteria as mentioned earlier: first, the object is not considered in its functional purpose. Second, the object is taken as a particular entity, individualized in other words. This impression is paradoxically still stronger because, instead of a pebble or of a shell, collected on a beach which would be the only one of its species, this industrial object is a mass product. By exhibiting it, the contrast is thus more striking. And therefore, the strangeness in which we eventually perceive it confers on its peculiarity an appearance—a subjective reality of its individual existence, which transforms it into a pseudo-person, that is to say in something that resonates with our emotional being.

So tipped over in 90 degrees, it appears in a different angle. Thus the curve of the urinal arouses a vague erotic evocation of the feminine body: the hips, the pelvis, the womb, the vagina… or of a masculine sex too. Not forgetting that venereal inspiration is always very active in the toilet, you can imagine Duchamp's propensity for wit on such a matter as a start. A white sanitary earthenware which we are currently used to see becomes a novelty. Third criterion: it would be hard to deny that the one who has the audaciousness to present such a thing commits himself. Especially as the current event of the First World War implies aggravating circumstances and makes this joke scandalous. That is why Duchamp can even be considered as a heroic artist. And on the whole, this might be a perfect example of an "auratic" piece of art.

The game is thus a kind of drift of the spirit, in which the result is not known. And for Duchamp it is exactly in this unexpected region that creation lies, in what he calls the "art coefficient."

> In the creative act, the artist goes from intention to realization through a chain of totally subjective reactions. His struggle toward the realization is a series of efforts, pains, satisfaction, refusals, decisions, which also cannot and must not be fully self-conscious, at least on the esthetic plane.
>
> The result of this struggle is a difference between the intention and its realization, a difference which the artist is not aware of.

Consequently, in the chain of reactions accompanying the creative act, a link is missing. This gap, representing the inability of the artist to express fully his intention, this difference between what he intended to realize and did realize, is the personal 'art coefficient' contained in the work. (Duchamp, 1994, pp. 188–189)

Marcel Duchamp's work corresponds well to the process described by Ehrenzweig who sees this work as a "new kind of cooperation between the artist and his public" (Ehrenzweig, pp. 137–139). But if modernity, directs creation to be an "open work," to use the term invented by Umberto Eco (1989), the nature of its process for the artist has probably never changed. Is not any work open, up to a point, because the spectator sees it through his personal prism of determinations by projecting his subjectivity into it?

Constraints, Rules, System

If creation is a game, as seen above, it raises the question of the rules. Rosalind Krauss says that Duchamp was strongly influenced by Raymond Roussel, fascinating character and wealthy dandy living in a kind of caravan which he had fit out luxuriously. In 1911, with Apollinaire, Francis Picabia and his wife, Duchamp attends a performance of Roussel's *Impressions of Africa*, in which are staged art-making machines: for painting, music, tapestry…

In his text entitled *How I wrote some of my books*, published in 1935, after his likely suicide, Roussel comments on his process which he thinks can be beneficial to "writers of the future" Without getting into details, it consists in issuing two almost identical sentences the words of which have a double meaning, and in using the one at the beginning, and the other at the end.

Both sentences being found, it was a question of writing a tale which can begin with the first one and finish with second. Thus it was from the resolution of this problem that I drew all my materials. (Roussel, 1977, p. 12)

Rosalind Krauss sees there the very principle of Duchampian creation: the ready-made is like the result of the machines in *Impressions d'Afrique* and Duchamp was been transformed into a kind of mechanical switch activating the impersonal production process of art. Thus, for Krauss (1997), to be an artist just becomes speculative activity by raising questions.

It seems to me that this type of interpretation totally misses what constitutes the pith and marrow of the process of creation. It is a typical "retrospective" point of view, as explained above which reduces the work to a linear route and inevitably reduces its complexity. It is even an almost hagiographical vision of art history, whereas one might consider that the work of Duchamp is as much a success as a failure. The technical procedure does not make the artistic process. Roussel is quite clear on this matter:

> This procedure, as a matter of fact, is similar to the rhyme. In both cases an unforeseen creation happens due to phonetic combinations. It is essentially a poetic procedure. Still it is necessary to know how to use it. And as well as with rhymes you can make good or bad verses, you can with such a procedure make good and bad works. (Roussel, 1977, p. 23)

Contrary to the game which is indeterminate, the rules are strict. Their aim is to fix, to constrain, to close. But paradoxically to define an open space, it is necessary to bound it. To acknowledge a movement, a fixed point is required. And to play, you need rules.

The idea is not new. Roussel reminds us. Poets have always used rhyme. Issuing a word you just have to find the rhyme. It is like hopscotch; you throw the pebble and it marks the place to which you have to skip. There is also the metrics: the number of feet which dictates the length of the verse to be made. The rhythmic brings it to complexity, in Latin verses for instance with dactyls and spondees… This has always been codified in versification rules. Above that, there still is the search for assonances and alliterations. Of course you can free yourselves from all that. You can even, like Baudelaire, make *Little poems in prose*. It might be still more difficult.

The rhyme is thus a "machine" to make poetry and even to make something new out of something old. Victor Hugo (1955) uses it and abuses it sometimes, not hesitating, for instance, to invent the city name of "Jerimadeth" to arrange a required rhyme. The parallel is obvious with the drawings of the same Victor Hugo. The "blot method" is not new. English painter Alexander Cozens had even formalized it as a teaching method of drawing in the previous century. Two hundred years before Leonardo recommended it to young painters if lacking inspiration.

> I will not forget to insert into these rules, a new theoretical invention for knowledge's sake, which, although it seems of little import and good for a laugh, is nonetheless, of great utility in bringing out the creativity in some of these inventions. This is the case if you cast your glance on any walls dirty with such stains or walls made up of rock formations of different types. If you have to invent some scenes, you will be able to discover them there in diverse forms, in diverse landscapes, adorned with mountains, rivers, rocks, trees, extensive plains, valleys, and hills. You can even see different battle scenes and movements made up of unusual figures, faces with strange expressions, and myriad things which you can transform into a complete and proper form constituting part of similar walls and rocks. These are like the sound of bells, in whose tolling, you hear names and words that your imagination conjures up. (Da Vinci, 1987, p. 247)

Max Ernst will "put Leonardo's lesson into practice in a very broad manner" systematizing it as a rubbing procedure in his *Frottages*. It lies on a phenomenon that is not however a surrealist invention though but it was already used by the painters of Lascaux Caves, 15,000 BC, inspired as they were by the hollows and the bumps of the rock the suggestive shadows of which already imposed to them some image of an animal.

Victor Hugo's blot of coffee or ink directs the poet-draughtsman he is on a graphic track, even if it usually ends in passionate waves or in a Germanic *Burg* at the very top of a sharp mountain, like a Walt Disney castle! Did not he also say that his poetical inspiration came to him more easily when taking a sheet of paper that had been crumpled before by his daughter Léopoldine? The page would then become less intimidating and more welcoming to his pen.

Victor Hugo is even said having played it with "holorhymes," that is to say two verses with the same pronunciation but two different meanings. One often attributed to him, for example, is: "Gall, amant de la Reine, alla, tour magnanime / Galamment de l'arène à la tour Magne, à Nîmes." (Gall, the Queen's lover, went, on a magnanimous tour / Gallantly from the Arena to the Magne Tower in Nîmes). Thus, besides the rhyme at the end of the verse, all the feet keep the rhyme, pacing and marching together!

All this has been since systematized. OULIPO made a method out of it. This "OUvroir de LIttérature POtentielle," that one could translate as Laboratory for Potential Literature, is a group of writers around poet Raymond Queneau and mathematician François Le Lionnais, who set systematic techniques of combinatory writing. The purpose of this method though is not the respect for the rules; it is to make them produce lucky finds. Raymond Queneau's inventiveness in his *Exercices de style*, for example, is due to the fact that the story being told has no interest by itself but by various ways it is told. All the intensity of Georges Perec's novel, *La disparition* (The Disappearance), lies in the fact that he exaggeratedly challenged himself to avoid the "e" letter, the most frequent one in French. In 1980, around Le Lionnais, was established a new branch for potential painting, OuPeinPo, OUvroir de PEINtiure POtentielle applying to painting the same sort of procedures (see Foulc, 2001).

Musicians too know very well how rules can be beneficial to them: from Beethoven taking as a starting point a rather insipid little waltz of Diabelli to compose his thirty three sublime variations, to the jazzman whose improvisation is sustained by a well-defined rhythm with even obligatory passages.

Often to let themselves be guided, the artists set the rules. Any art teacher knows the educational virtues of a constraint: I remember wonderful works of pupils in a nursery school, who had been asked to express the jubilation of spring which had just arrived only by the means of big sheets of white paper and an enormous can of black paint.

Examples are uncountable. English sculptor Tony Cragg refrained to stick, to nail, to screw the elements he uses, whereas his work is a matter of assemblage. Withdrawing from an intentional gesture and skill, French painter Simon Hantaï set himself "the folding method," plunging his folded canvasses in paint, the work being the result of mechanical operations. Another French artist, François Morellet, expresses very well this attitude:

> For me, a "system", it is a kind of very concise rules of the game which exist before the work and determines exactly its development and thus its execution. I chose this term because it could indicate an attitude I like very much, that of the artists who do not identify with what they are doing [...]. The system allows to decrease the number of subjective decisions and to let the work progress by itself so to speak in front of the spectator. (Morellet, 2000–2001)

The term "system" is of course to be understood in the positive sense of a method to find something and not in the pejorative meaning of a mechanical automatism where everything is played beforehand. In many artistic means such as etching, clay modeling, photography, the medium itself sets up inescapable technical constraints. One could even think that the more the constraint, the less we know where it will lead and thus the more there is room for creation, the "art coefficient," as Duchamp would have said.

The End—When Is a Work Finished?

In French the word *réussir* (to succeed) comes from Italian *ri-uscire*, which initially only means that an exit has been found. This relativization of success, or at least its uncertainty for the artist, in real time, while working, raises the question of the end. Concerning official painters who claimed that Whistler did not finish his paintings, Whistler had this scathing reply: "Their works may be finished but they have certainly not been started" (Chaleyssin, 1995b, p. 141). The most important is at first that there is some substance, that the work "kneads" a problem.

But what is thus a finished work? Whistler proposes an interesting criterion:

> A picture is finished when all trace of the means used to bring about the end has disappeared. (Chaleyssin, 1995a, p. 144)

We could say that this disappearance of all trace of the means corresponds to the withdrawal of all what the artist had put in it too deliberately. Moreover, Braque (1952) echoes him, saying that "The picture is finished when it has erased the idea" (p. 27). Exaggerating it we could almost say that the work is finished, when the artist was able to give up the purpose and the end that was first expected. Thus logically it is necessary to aim for this incompletion. The German painter Emil Schumacher (1997) claimed that: "A painting is never finished. It must not be finished. Any finished painting is also finished in the worst sense of the word."

And Picasso again:

> I go very slowly. I do not want to spoil the first freshness of my work… If it were possible for me, I would leave it as it is, even if it means beginning again and bringing it to a more advanced state on another canvas. Then I would act the same with this one… There would never be a "finished" canvas, but various "states" of the same picture which usually disappear during the work… Do not finish or execute have moreover a double meaning? To end, to finish, but also to kill, to give the deathblow? (Picasso, p. 108).

Perhaps could we just say that a work is "finished enough" when something material and perceptible has embodied something that is intelligible. A finished work is again accompanied by a certain disorder.

How to Be an Artist?

To this question I have been trying to reply for over thirty years, as an artist myself first, but not in the sense of an acknowledgement as an artist, in the way George Dickie approaches it on an almost sociological perspective. For Dickie (1984), the one who is involved in creation applies for recognition. He/she is a candidate artist and it is the Art Circle that decides to confer (or not) this recognition. In my personal case I am a worldwide unknown artist. Nevertheless artistic creation is central in my existence. Thus the answer which I try to bring to this question is, on the "poietic" side. I try to know what characterizes an artistic attitude and process? And this, regardless of the fact that it is recognized (or not).

In this endeavor, I draw on my personal artistic experience—from what else could I start? —and in my teaching practice, to try to accompany young people building their own artistic process but also developing themselves as individuals.

Some general principles have been given here, but there are just general principles. There are so many ways to engage in the artistic process. You cannot really teach it. Everyone has to find his or her way and art teaching tries to give students the conditions that could make their creation possible. Of course there are little tricks that could facilitate it but if they can help "creativity," it has little to do with "creation."

In the many interviews I made with artists (Toulouse & Molina, 2012) they all say that, even if they received an art education, it is not in their school that they had learned to be artists. There is always a claim of self-education. At school, they found a context with other artists. They may have been introduced to the "Art Circle." But it is by themselves, on their own, that they have become artists. To teach art is thus a very personalized concern, trying to bring every single student to become aware of what is happening, and beyond main principles as those which are expressed in this chapter, it is very difficult to give recipes which would surely work, because as Jean Dubuffet puts it: "Real art is always where you don't wait for it (Dubuffet, 1967, p. 201)."

Win or Lose

Success is relative in art. For Giacometti, success cannot exist without failure and often "failures" turn out later to be "successes." The criterion of success is difficult to establish. A work with which you can be very dissatisfied at the end of day can appear the next day full of interest. Conversely, your success in the studio can seem very poor the day after, when excitement has dropped off, and can even look so unbearable that you want to destroy it absolutely to remove the kind of shame which it arouses for you, like a cat hides its' mess. But one has to be cautious with inconvenient fits of temper, whatever direction they might have. Very often the relevant assessment criterion for the work can only be described afterward. We cannot judge a current work as we would a validated outcome and every stage must only be judged by the capacity which it provides to go on. From this point of view even a recognized failure is a good experience if it allows one to understand what was bad. Picasso clarifies this further when he tells Sabartès:

> In museums, for example, there are only failed paintings... You laugh; pay special attention to know if I am right; what we take now for masterpieces, it is what went the furthest from the rules dictated by the masters of the time. The best ones reveal most clearly the stigmas of the artist who painted them. (Picasso, 1998, p. 166)

This is moreover the story of the *Young Ladies of Avignon*. This inaugural picture for cubism is rightly considered as the kick-off of modern art at the dawn of the twentieth century. When, after a large number of drawings and exploratory studies, this picture is created, between May and July 1907, at the "Bateau-Lavoir" in Montmartre, there is a general consternation. Picasso's friends, the future masters of the time, pioneers of painting themselves, feel desperately confused. Georges Braque says that it is as if one was "drinking some oil or eating burning tow." Derain dreads that Picasso would be found "having hanged himself behind his picture." Matisse gets angry and speaks of "giving up" on Picasso. The usually enthusiastic critic Félix Fénéon, the art-dealer Ambroise Vollard, the collector Gertrude Stein are distressed and shocked. Leo, the brother of the latter, who had considered two years before Picasso "as a first-grade genius and one of the best alive draughtsmen," even talks about an "abominable formless waste." A few years later, André Salmon, to whom the title of the picture is due writes:

The regular visitors at this curious studio in Ravignan street trusted the young master and they were generally disappointed when he allowed them to judge the first state of its new work [...]. It is the hideousness of the faces that froze those half-converts with dismay. (Daix, 1977, p. 90)

Nevertheless the picture seems to have moved with the painter through his various studios, still stretched on its frame. In 1916 it is shown in a confidential exhibition. In 1918 when Picasso settles down with Olga, in a *bourgeois* flat of the fashionable La Boétie street, the canvas is then unstretched and rolled in a corner of the studio. In 1921, André Breton and Aragon will be the middlemen for its purchase by the dressmaker Jacques Doucet. The review *La Révolution surréaliste* will publish in 1925 its first photographic reproduction which will make it known. But it is only in 1937, after having been sold by Doucet's widow to the Seligmann Gallery that it will be really shown, in New York, and in 1939 definitively acquired by the MOMA.

In spite of his legendary self-confidence, we cannot think that Picasso was not a little shaken himself by this story and, if he had not been able to convince his closest companions of the validity of this painting, it may be because he was not totally sure himself. Because contrary to the incomprehension expressed to the impressionists, the hostility did not come from conservative circles but from what we could call the artistic *avant-garde* of the moment.

What Could Research on Creation Be?

I participated in the CREAPRO, creative process research program the aim of which was to try to compare the creative processes in various fields (art, musical composition, scenario writing, design, and engineering). To begin, I was quite interested in such a comparison but very soon I realized that the methodology of this project did not fit with what I was trying to approach. If the methodology could work (more or less) for other subjects than mine, it was absolutely not adapted to art where there is no bill of specifications imposed to the artist who does not "execute" his work. Most of the time, even when commissioned, the artist has not to provide something precisely defined at first. The frame of creation is much more open for an artist than for a designer, a scenario writer and still more than for an engineer.

Initiated by a team of researchers in psychology, the scenario of this project was conceived in three sets. First an inquiry among "experts" was supposed to allow the typical profile of a creator to be described. I supervised interviews with thirty artists and this was probably the most interesting part of the project. We made a book and films out of them (Toulouse & Molina, 2012). We were in charge of the art part and we conceived an interview grid. The treatment of those interviews by our partners was then strictly quantitative, measuring the statistical occurrence of the used words with a lexical analysis application called "Tropes." But the problem is that artists can speak of the same thing in different words or of something different with the same ones. But above that, such an exploitation of data seemed to us totally irrelevant because creation probably is an exacerbated experience of singularity and a strictly statistical study would miss the object the research is deemed to approach.

The second phase consisted of observing "novices" (that is to say art students) at work. We had chosen a group of students working on etching which is a very interesting medium to observe a creation process because the medium is very technical and because of the many surprises that happen both when etching the plates as when printing. The aim was to see whether the template of the typical creator could be seen among them, making a difference between the more successful ones and the others. I suggested it was perhaps more interesting for the enquiry and for the students to try self-observation. I asked them to hold a diary, (as many artists do) writing week after week how things had gone on and comparing it afterward with what they had first thought and this document was very useful to discuss and to help them to find their own way in creation. Our psychologist partners wanted them to fill a form at the end of each session to define the tasks they had been working at on that day. I made a grid for that thinking so it could shed light on their process. This could have been quite interesting but, here again, we did not agree with the kind of processing our partners made with the collected information. Instead of analysing every single case to understand the specific route of each student, case by case, the research team examined general trends (see Chapter 3 in this volume). However, for us the only relevant approach was case study methods, without any guarantee to be really able to make even a typology, and almost avoiding to try to statisticize anything. Just to map different routes!

The third part of the program aimed to conceive a remediation to improve creative abilities of those who had difficulties. To tell the truth, a long practice of art education has allowed us to sort out two or three principles which always have to be adapted to specific situations and to particular persons, but they can be summarized in a simple precept: to relativize one's conscious will that would inevitably lead to what one already knows; to let then outer determinations work by themselves and issue other possible combinations which one never would have thought of and among which there shall be much more interesting things to choose. As explained above, the aim thus is to train the students at a mental seesaw between receptiveness and action, the most difficult task not being to make them produce interesting things, but to make them become aware that they are.

Moreover our participation in this research program has been a very instructive experience which permitted us, on the contrary, to define better our positions and to argue things that had seemed obvious to us, but that were in fact only prejudices or just hypotheses. And does not being a researcher imply a critical attitude on methods? It is not the least part of research.

Put briefly, our criticism of the methodology of this program draws on the specificity of our research topic. An artist is not outside his or her process, just observing it. He or she is intimately and emotionally involved in the creation and so is an artist-researcher: he is not only a food critic, he or she has to be a cook (and tries to be a chef), he/she is not only a theologian but a mystic, he/she is not only a sexologist or a kama sutra technician but a lover.

Conclusion

Art education is probably the very paradigm of any education. An educational work is necessarily individualized and our practice makes us think that there is no miracle solution. It is of course by getting lost that Christopher Columbus discovered America, but it is not enough to get lost to find a New World. It is probably because he was in the grips of despair that Van Gogh was a great painter, but it is not enough to be unfortunate to be a creator… Creation, by definition, presents a character of exception. From Leonardo's notebooks to Matisse's remarks on art, through Delacroix's diary, and without forgetting the uncountable texts of older or contemporary artists—often entitled "Conversation with…", there is already an immense corpus to study.

The path of a creation is always singular, because to create, it is, every time, to make something appear that had been never heard, never seen before. It is hardly quantifiable. Contrary to usual statistical studies, you cannot consider as unimportant infinitely minor occurrences because they are precisely the very ones that become the source of creativity. It is a general epistemological problem. Likewise, the possibility of butterflies having wings with the same design and the same color than the leaves of the trees on which they stay was infinitely improbable in the evolution of species. And nevertheless, they are the only ones who escaped predators, survived and who, then, by genetic mechanisms, reproduced in a perfect statistical determinism. In the same way, if we can doubtless highlight "creative" features common to numerous activities, the reproductible character of a creation cannot itself be presumed.

Bibliography

Artaud, A. (1968). *Le Pèse-nerfs* dans *L'ombilic des limbes*. Paris: Gallimard, "Poésie".

Bacon, F. (1996). *Entretiens avec Michel Archambaud*. Paris: éd. Gallimard, coll. "Folio".

Barthes, R. (1989). *Leçon*. Paris: Seuil, coll. "Points".

Baudelaire, C. (1983). *Curiosités esthétiques, L'art romantique*. Paris: Garnier, coll. Classiques.

Benjamin, W. (2008). L'œuvre d'art à l'époque de sa reproductibilité technique, (1939), in "*Œuvres*", T. 3. Paris: Gallimard.

Bergson, H. (1970). *Œuvres*. Paris: PUF.

Braque, G. *Le jour et la nuit. Cahiers 1917–1952*. Paris: Gallimard.

Brassaï. (1969). *Conversations avec Picasso*. Paris: Gallimard, coll. "Idées".

Canaës, M.-H., & Marchand-Zanartu, N. (2011). *Images de pensée*. Paris: RMN.

Chaleyssin, P. (1995a). *James Mc Neill Whistler, Le Cri strident du papillon*. Bournemouth: Parkstow.

Chaleyssin, P. (1995b). *John McNeil Whistler*. Paris: Ed. de l'Olympe.

Charbonnier, G. (1980). *Le monologue du peintre*. Neuilly-sur-Seine: Guy Durier.

Cheng, F. (1991). *Vide et plein, le langage pictural chinois*. Paris: éd. Seuil, coll. Points.

Da Vinci, L. (1987). Les carnets, *Tome 2*. Paris: Gallimard, coll. "Tel".

Daix, P. (1977). *La Vie de peintre de Pablo Picasso*. Paris: Seuil, coll. "Points".

Delacroix, E. (1980). *Journal*, le 15 mai 1824. Paris: Plon.

Deleuze, G. (1971). *Proust et les signes*. Paris: PUF, coll. "À la pensée".

Deleuze, G. (1996). *Bacon - Logique de la sensation*. Paris: La différence.

Derrida, J. (1978). *La vérité en peinture*. Paris: Flammarion, coll. "Champs".

Dickie, G. (1984). *The art circle*. New York: Haven.

Dubuffet, J. (1967). *Prospectus et tous écrits suivants, tome 1, réunis et présentés par Hubert Damisch*. Paris: Gallimard.

Duchamp, M. (1994). *Duchamp du signe*. Paris: Flammarion, coll. "Champs".

Eco, U. (1989). *The open work*. Cambridge: Harvard University Press.

Edwards, B. (1979). *Dessiner grâce au cerveau droit*. Liège: Mardaga.

Ehrenzweig, A. (1991). *L'ordre caché de l'art, essai sur la psychologie de l'imagination créatrice*. Paris: Gallimard, coll. "Tel".

Emmerling, L. (2003). *Pollock*. Köln: éd. Taschen.

Forcellino, A. (2006). *Michel-Ange, une vie inquiète*. Paris: Seuil.

Foulc, T. (2001, mai). *Vingt ans de peinture potentielle* dans le *Magazine littéraire*, no. 398.

Freud, S. (1976). *L'interprétation des rêves*. Paris: PUF.

Gautier, T. (1954). *Poésies complètes*. Paris: éd. Garnier, coll. "Classiques".

Giacometti, A. (1990). *Écrits*. Paris: Hermann.

Hadamard, J. (2007). *Essai sur la psychologie de l'invention dans le domaine mathématique*, suivi de Poincaré Henri, *L'invention mathématique*. Paris: Jacques Gabay.

Harrison, C., & Wood, P. (1992). *Art en théorie 1900–1990*. Paris: Hazan.

Hartt, F. (1987). *Le David de Michel-Ange, le modèle original retrouvé*. Paris: Gallimard.

Hogarth, W. (1753). *Analysis of beauty*. http://www.tristramshandyweb.it/e-texts/hogarth/analysis_html.

Hugo, V. (1955). *La légende des siècles*. Paris: Gallimard, coll. "La Pléiade".

Juan De Mendoza, J.-L. (1996). *Deux hémisphère, un cerveau*. Paris: Flammarion, coll. "Dominos".

Juliet, C. (1978). *Rencontres avec Bram Van Velde*. Fontfroide le Haut: Fata Morgana.

Juliet, C. (1990). *Entretiens avec Pierre Soulages*. Paris: L'Échoppe.

Kandinsky. (1969). *Du spirituel dans l'art et dans la peinture en particulier* (1912), Paris: Denoël.

Klee, P. (1973). *La pensée créatrice*. Paris: Dessain et Tolra.

Krauss, R. (1997). *Passages, Une histoire de la sculpture moderne de Rodin à Smithson*. Paris: Macula.

Lemoine, S. (1986). *Dada*. Paris: Hazan.

Lévi-Strauss, C. (1986). *La pensée sauvage*. Paris: Plon, coll. Pocket.

Malifaud, P. (1997). *Métaphysique, Tentative de dévoilement du réel*. Paris: Hommes & Groupes.

Malifaud, P., & Toulouse, I. (2012). *Décoder le réel, Dialogue*. Paris: L'Harmattan, coll. "Eurêka & Cie".

Marx, K. (1963). *Avant-propos à la Critique de l'économie politique* (1859) dans *Œuvres*. Paris: Gallimard, coll. "La Pléiade", T. 1.

Matisse, H. (1972). *Écrits et propos sur l'art*. Paris: Hermann, coll. "Savoir".

Matisse, H., Couturier, M.-A., & Rayssiguier, L.-B. (1993). *La chapelle de Vence, journal d'une création*. Paris: Coédition Cerf, Menil Foundation, Skira.

Merleau-Ponty, M. (1983). *Le visible et l'invisible*. Paris: Gallimard, coll. "Tel".

Millet, C. (1988). *L'art contemporain en France*. Paris: Flammarion.

Morellet, F. (2000–2001). *Catalogue de l'exposition*. Paris: Galerie nationale du Jeu de Paume.

Panofsky, E. (1975). *La perspective comme forme symbolique*, traduction sous la direction de Guy Ballangé. Paris: Minuit, coll. "le sens commun".

Panofsky, E. (1988). *Les antécédents idéologiques de la calandre Rolls Royce*. Paris: Le promeneur/Quai Voltaire.

Passeron, R. (1992). *L'œuvre picturale et les fonctions de l'apparence* (3e édition). Paris: Vrin.

Passeron, R. (1993). Pour une approche poïétique de la création. In *Les Enjeux*, T. 1, Paris: Encyclopaedia Universalis.

Pevsner, N. (1956). *The Englishness of English art*. London: Penguin.

Picasso, P. (1998). *Propos sur l'art*, Réunis par Marie-Laure Bernadac et Michael Androula. Paris: Gallimard.

Poincaré, H. (2003). *La valeur de la science*. Paris: Flammarion.

Préparation Militaire Supérieure. (1939). *Manuel, T. 2, Infanterie 1ʳᵉ et 2ᵉ années.* Nancy, Paris, Strasbourg: Berger-Levrault.

Proust, M. (1971). *Contre Sainte Beuve,* précédé de *Pastiches et mélanges* et suivi de *Essais et articles.* Paris: Gallimard, coll. "La Pléiade".

Rothko, M. (2007). *Écrits sur l'art, 1934–1969.* Paris: Flammarion.

Roussel, R. (1977). *Comment j'ai écrit certains de mes livres.* Paris: UGE, coll. "10/18".

Schumacher, E. (1997). *Catalogue de l'exposition.* Paris: Galerie Nationale du Jeu de Paume.

Seckel, H. (1988). *Catalogue de l'exposition. Les demoiselles d'Avignon.* Paris: Musée Picasso.

Soulages, P. (1961). "Lettre ouverte no. 3", *De refus en refus,* recueil de textes réunis par Raymond Girard, Paris.

Toulouse, I. (direction) (2008). *Euréka, le moment de l'invention.* Paris: L'Harmattan.

Toulouse, I. (2012). *Clair-obscur - Essai sur la pensée créatrice.* Paris: L'Harmattan, coll. "Euréka & Cie".

Toulouse, I., & Molina, M. A. (2012). *Théories de la pratique, - Ce qu'en disent les artistes.* Paris: L'Harmattan, coll. "Euréka & Cie".

Valéry, P. (1957). *Œuvres, T. 1.* Paris: Gallimard, coll. "La Pléiade".

Valéry, P. (1972). *Introduction à la méthode de Léonard de Vinci.* Paris: Gallimard, coll. "Idées".

Valéry, P. (2003). *Degas, danse, dessin.* Paris: Folio, coll. "Essais".

Van Gogh, V. (1953). *Lettres à son frère Théo.* Paris: Gallimard.

Varenne, J. (1989). *Aux sources du yoga.* Paris: Jacqueline Renard.

Vermersch, P. (2003). *L'Entretien d'explicitation.* Issy-les-Moulineaux: ESF.

Wittkower, R. (1995). *Qu'est ce que la sculpture? Principes et procédures de l'Antiquité au XXe siècle.* Paris: Macula.

3

The Creative Process in Graphic Art

Marion Botella

The visual arts have traditionally been treated as a major domain of creative activity and were the subject of some of the first empirical studies of creativity, such as Patrick's seminal research (Patrick, 1935, 1937). Artists are considered as an archetype of creators (Schlewitt-Haynes, Earthman, & Burns, 2002; Stanko-Kaczmarek, 2012). In this chapter, the artistic creative process will be examined through: (1) interviews of professional artists, (2) observations of art students in real contexts, and (3) exercises proposed to art students to develop their creative process; the creative process will be examined with a particular attention paid to the factors involved in artistic creativity using the multivariate approach to creativity. This approach describes four main categories of factors: cognitive, conative, emotional, and environmental factors (Lubart, Mouchiroud, Tordjman, & Zenasni, 2015).

M. Botella (✉)
Paris Descartes University, Paris, France
e-mail: marion.botella@parisdescartes.fr

T. Lubart (ed.), *The Creative Process*, Palgrave Studies in Creativity and Culture,
https://doi.org/10.1057/978-1-137-50563-7_3

The Creative Process in Art

Since Wallas (1926), many authors have tried to determine the stages describing the creative process (Busse & Mansfield, 1980; Osborn, 1963; Treffinger 1995), using analyses of historical documents (Gruber, 1981), interviews (Shaw & Runco, 1994), observational booklets (Feist, 1994), or brain imaging (Arden, Chavez, Grazioplene, & Jung, 2010; Dietrich & Kanso, 2010). These conceptions concern creativity, in general, including art, science, design, or music (for a review, see Botella, Nelson & Zenasni, 2016); other conceptions concern specific creativity only for one domain. Models describing specifically the artistic creative process are rare but exist.

Mace and Ward (2002) proposed a specific model of the artistic process based on interviews with professional artists; it is a dynamic model in four stages. The artistic process begins with the *conception* of the artistic work. The work is introduced by an idea or a more or less vague impression. The second stage corresponds to the *development* of the idea. Artists structure, complete, and restructure the idea. Also, they identify the work development possibilities according to their ideas and feelings. Furthermore, artists make implicit and explicit decisions by observing their work. This evaluation incites them to question the ideas, expressions, metaphors, and analogies that they wish to use and those that they prefer to abandon or to put aside for future work. The third stage is the *realization* of the idea in which artists transform the idea into a physical entity. The fourth and the last stage is the *finalization* and the resolution of the artistic work. The artists evaluate the production: they can choose to end the production, to pursue it, to abandon it, to postpone it, to store it, or to destroy it. If the artists consider that the production is a success and satisfying, they can choose to expose it.

Whatever is the result of the artwork, artists constantly enrich their experience and their knowledge. Their knowledge is the result of a dynamic and perpetual interaction with artistic practice. The artists add and refine their skills, techniques, and knowledge. Also, they sharpen their interests and their artistic personality. During a work, new ideas can appear and be reused later in a new work of art. Thus, Mace and

Ward (2002) did not propose a linear but rather a dynamic and iterative model as far as the artistic process is under the constant influence of multiple factors, including the development of other productions.

Getzels and Csikszentmihalyi (1976) found that artistic creativity is related to time spent in an exploratory phase before starting to draw. In a field study of ink painters, Yokochi and Okada (2005) observed that the painter formed a global picture with each successive element. The painter had a partial image in his or her head and each line drawn constrained other lines. In this way, ink painting seemed to be a set of many successive pictures in which each picture engaged its own art process.

Based on Mace and Ward's model in art and the general literature on the creative process, we examined the dynamic creative process and its association with affect in the graphic art field (Botella, Zenasni, & Lubart, 2011a) on two samples of art students. Nine stages of the artistic creative process were considered: preparation ("I collect information or I reflect about the subject"), concentration ("I concentrate on the work to be realized"), incubation ("I let my ideas flow alone"), ideation ("I think of new ideas"), insight ("Suddenly, I know what I am going to do"), verification ("I check my ideas"), planning ("I plan my work"), production ("I realize/compose my ideas"), and validation ("I verify if my work is done"). These stages were linked to three measures of affect: positive mood, negative mood, and arousal. Results indicated that positive mood increased during the creative process, whereas negative mood decreased. Additionally, preparation and ideation stages were reported to be more negative than other stages: the preparation stage is the beginning of the process, and the art students do not know yet what they will do; for ideation, they have new ideas, but they do not know how to use them. This could generate stress. Stages of concentration and incubation are not necessarily pleasant but not as negative as preparation. Having an insight is a turning point in the artistic creative process, associated with positive mood and high arousal; the art students know then what they want to do. The production stage is relatively positive even if some negative mood is reported for some students. Finally, validation is the most positive stage because it is the ending of all the work done and it could result in a feeling of freedom. This study employed a dynamic approach to an artistic creative process including affects.

As Fürst, Ghisletta, and Lubart (2012) who examined personality, cognition and affect in the creative process of visual art students, or Glăveanu (2013) who considered the "doing" stages of the creative process according to knowledge, procedures, undergoing material conditions, undergoing social context and obstacle of folk art creativity, today, the cognitive dimension is not the only one studied in creativity. It is important to consider all the multivariate factors (cognitive, conative, affective, and environmental) involved in the creative process to improve its understanding.

The Multivariate Approach

The profile of creative people, artists in particular, has been extensively studied (Batey & Furnham, 2006; Feist, 1998; Furnham, Batey, Booth, Patel, & Lozinskaya, 2011). However, the cognitive, conative, emotional and environmental characteristics of creators are mainly studied separately. According to the multivariate approach, the creative person is defined by a particular combination of cognitive, conative and emotional dimensions associated with favorable environmental conditions (Amabile, 1983, 1996; Gardner, 1993; Lubart, 1999; Lubart et al., 2015; Sternberg & Lubart, 1991, 1995).

The *cognitive dimension* corresponds to the intellectual abilities involved in creativity. Lubart et al. (2015) propose a summary of cognitive capacities including synthetic capacities of identification, definition, and redefinition of the problem. Also, selective encoding permits the selection of relevant information for solving the problem. The selective comparison ability helps one to observe similarities between various domains. In addition, the creative person makes associations between the ideas collected (selective combination). Several authors have proposed an elaboration–evaluation cycle in which ideas are perpetually generated and judged (see Bonnardel, 1999). These cognitive capacities favor the emergence of a creative solution (Lefebvre, Reader, & Sol, 2013; Sternberg & O'Hara, 2000).

The *conative dimension* concerns personality traits and motivation. Creative individuals are usually described as open to new experiences

(Barron, 1969; Feist, 1998; Furnham & Bachtiar, 2008; Gough, 1979; Mac Kinnon, 1965; McCrae & Costa, 1987; Wolfradt & Pretz, 2001; Zenasni, Besançon, & Lubart, 2008). Openness is reflected in a dynamic fantasy life, aesthetic sensibility, emotional awareness, need for originality, intellectual curiosity, and a strong personal value system (Helson, 1999). Creative people are also tolerant of ambiguity (Barron & Harrington, 1981; Levy & Langer, 1999; Sternberg & Lubart, 1995; Tegano, 1990; Zenasni & Lubart, 2001, 2008). In a meta-analysis of personality traits for scientific and artistic creativity, Feist (1998) showed that artists are characterized by their openness to new experiences, fantasy and imagination, their lively, ambitious and nonconformist nature. Additionally, motivation is an important key to creativity (Nakamura & Csikszentmihalyi, 2003). Especially, Stanko-Kaczmarek (2012) showed that intrinsic motivation increases positive affect during the creative process of fine arts students, indicating that "cognitive and affective processes are related in the creative process" (p. 308).

The *emotional dimension* of the multivariate approach corresponds to emotional traits (Botella, Zenasni, & Lubart, 2011b, 2015) and states (Baas, De Dreu, & Nijstad, 2008; Zenasni & Lubart, 2008). For example, emotional clarity and the capacity to perceive feelings are positively correlated with creative performance (George & Zhou, 2002). Also, emotional intelligence increases generosity and vigor, favoring creativity outputs (Carmeli, McKay, & Kaufman, 2014). Examining the link between emotional intelligence and creative personality, Wolfradt, Felfe, and Koster (2002) observed moderate to strong correlations between the two concepts (from $r=.36$ to $r=.55$, $p<.01$). In artistic creativity, Feist (1998) indicated that artists, compared to scientists, tend to be more emotional, anxious, emotionally unstable, and to have a strong emotional sensitivity.

Finally, the multivariate approach emphasizes the *environment* which offers physical and/or social stimulations and can help the generation and maturation of ideas, thus reinforcing motivation (Lubart, 1999). The environment includes the appreciation of creativity through social judgement or by learning to judge creativity (Storme, Myszkowski, Çelik & Lubart, 2014). Moreover, according to Bilton (2013), the environment of visual artists includes vertical (galleries, critics, collectors) and horizontal networks (friends, fellow-artists).

For Sternberg and Lubart (1995), creativity involves more than a sum of all these components: certain constituents can partially compensate each other. For example, a strong degree of motivation can mitigate a lack of knowledge. These components interact among themselves; the combination of high intelligence and strong motivation may enhance creative performance in a multiplicative way. Thus, the multivariate approach focuses attention on the various constituents involved in artistic creative activity and aims to examine the interactions between them.

The multivariate approach was mainly used to describe the characteristics of creative individuals. The objective of this chapter is to use it to describe the creative process of artists and art students. First, we will see how professional artists describe their own creative process through interviews. Then, the creative process of art students will be self-reported during a real context of creation. Finally, two exercises will be proposed to art students in order to enhance their creative process.

The Creative Process According to Professional Artists

The purpose of this section is to examine the factors engaged in artistic creativity and to describe the creative process based on artists' narrative accounts of their work. In particular, an interview-based study of 27 professional artists (18 men and 9 women, $m = 46.36$ years; $s = 8.71$; *range* $= 30$–66 years) will be described (Botella et al., 2013).

The analysis is based on the spontaneous discourse of artists with regard to the cognitive, conative, emotional, and environmental aspects that can influence their creativity and, also, the stages of their artistic activity. Interviews were associated with an auto-questionnaire assessing 39 multivariate factors potentially involved in artistic creativity.

The Stages of the Creative Process

A qualitative coding is applied to extract the stages of the creative process from the artists' discourse. This kind of analysis depends, to a large extent, on the skills and appreciation of the analyst, while being firmly grounded in the narrative material under investigation. The construction of a coding frame is therefore, at the same time, data- and theory-driven. Overall, art-making activity seems to involve six distinguishable stages according to the artists' discourse:

1. The creative process begins with *an idea or a "vision."* This first creative idea comes after a period of void, of wandering and could be triggered by an image, a sight, a sound that resonates with what had "matured" inside the artist for a long time. These initial ideas are general in nature and they become more "specific" or "concrete" during the next stages.
2. A second phase is *documentation and reflection.* Most visions are incomplete and need careful consideration that amounts to an "incubation" phase. At this stage, artists sometimes need to gather more information about the materials and technologies required in order to turn their vision into reality.
3. The third moment is represented by the *first sketches,* first attempts at giving the project a material form. This can overlap with documentation or be postponed depending on how the artist works. Some like to immediately draw their project, some wait to make a "model," such as a clay version that requires more elaboration.
4. After sketches, there is a more or less extended period of *testing the forms and ideas* that originated from reflection and preliminary work. Missing adequate tools or materials is a common difficulty encountered at this stage. Also, forms of material undergoing come to the fore.
5. These tests, of making and remaking, end up in *provisional objects,* "drafts" and almost-finished products. This stage requires working on the details of the work, perfecting its features, adjusting it to the context of presentation. Undergoing the final result at this stage invites

evaluative judgements on the part of the artist, whereas some argue that artistic work is never "done," never complete, but there is a moment when it needs to end.
6. It was also often the case for a first object to be followed by *a series*. In this situation, variations can be made and the vision finally completed.

This description of the artistic creative process has the advantage of taking into account the real-life experience of the professional artists as described in their own words. Between the six phases of work, there are many feedback loops and a movement of "back and forth" which can be better captured through longitudinal observation.

The Multivariate Factors Involved in Artistic Creativity

Additionally, a semantic analysis is applied to extract the multivariate factors from the discourse. This kind of analysis uses software which includes a hundred thousand semantic classifications corresponding to predefined word classifications (concepts and related terms that are theoretically close). The semantic analysis allows extracting *facilitating factors* grouping terms which, according to the discourse of the artists, have a positive impact on their creativity. These factors involve cognitive dimensions (intelligence, knowledge or memory, etc.), conative dimensions (curiosity, spontaneity and sincerity, etc.), emotional dimensions (satisfaction, happiness, frustration, or anger, etc.) and environmental dimensions (space, luminosity, need to be alone, in a "bubble," etc.). *Inhibiting factors* referred mainly to the environment and features such as constraints, codes, rules, risks, finances, and deadlines. Cognitive dimensions (contradictions, illusions and uncertainties, etc.) and emotional dimensions (fear and guilt, etc.) can also inhibit the artistic creativity.

Nothing mentioned in the artists' discourse made reference to conative aspects as an inhibiting factor. Based on the number of words used for each multivariate factor, in terms of facilitating factors, the artists mentioned primarily three aspects (cognition, conation, and emotion). When the artists spoke about factors facilitating their creativity, they mentioned equally the cognitive, conative, and emotional aspects. However, when artists evoked factors inhibiting their creativity, they spoke mainly about emotional aspects. Cognitive and conative dimensions were essentially mentioned as being favorable to creative work, whereas the results were mixed for the emotional aspects.

Additionally, results from the questionnaire, assessing 39 multivariate factors potentially involved in artistic creativity, offer further information for the multivariate approach. The mean of all the cognitive dimensions appears to be the most important, more than the conative dimensions, the environment, and, finally, the emotional dimensions. "Selective combination" appears essential for artistic creativity (Getz & Lubart, 2000). Also, "self-criticism" allows beginning to estimate the work and to envision possibilities of improvement (Goor & Sommerfeld, 1975). In contrast, according to the artist's self-report, the social aspects were not essential for their creativity. In particular, sociability and extraversion were not seen as necessary for creation.

Observations of the Creative Process with Art Students

The purpose of this second section is to examine longitudinally the multivariate factors engaged in real-life settings of art students who create work as part of their master program courses. In the research that will be highlighted here (Botella & Lubart, 2015; Botella, Zenasni, Nelson, & Lubart, Forthcoming), 27 undergraduate art students at a French university participated (21 females, 6 males, $m = 22.75$ years, $sd = 1.16$ years, age range: 21–25 years). Based on interviews with artists described in the previous section (Botella et al., 2013) and other creators (Glăveanu et al., 2013), a booklet was constructed consisting of a structured self-report focused on stages of the creative process in which participants indicated their weekly progress. Thirteen stages of the creative process are presented in the booklet: definition of the problem, reflection, documentation, consideration of constraints, insight, associative thinking, divergent thinking, convergent thinking, the benefit from chance, implementation, finalization, judgement, and taking a break. In addition to selected stages of the creative process, students self-evaluate the use of a series of multivariate factors in their creative work: perseverance, rigor, patience, perfectionism, energy, tolerance for ambiguity, optimism, openness, intuition, risk-taking, satisfaction, surprise, stress, frustration, disappointment, to talk with others, to ask for an opinion, to express ideas, to persuade others, and team work. Art students have one semester—12 weeks—to create freely an artwork. At the end of each week, students had to complete a page of the booklet and indicate the stage(s) of the creative process and the multivariate factors they engaged during that session.

The Creative Process and the Multivariate Factors

An analysis of the transitions between stages is conducted on the 13 selected stages of the booklet and a mean multivariate profile (cognitive, conative, emotional, and environmental characteristics) is calculated for each stage of the artistic creative process on the overall sample. Figure 3.1 resumes the most frequent transitions between stages and the most important factors for each stage. It is important to note that all transitions are possible but only the most typical are presented.

The "*definition* of the problem" stage is frequently followed by documentation, insight, and chance. This stage is not associated with a specific multivariate profile. *Reflection* is typically followed by definition and judgement. This stage involves environmental factors in which art students talk to others. *Documentation* leads back to definition and to divergent thinking. It is stressful and frustrating for art students. The stage of *consideration of the constraints* is followed by finalization and involves risk-taking. Art students continue typically the stage of *insight* by defining the problem, documentation and benefiting from chance.

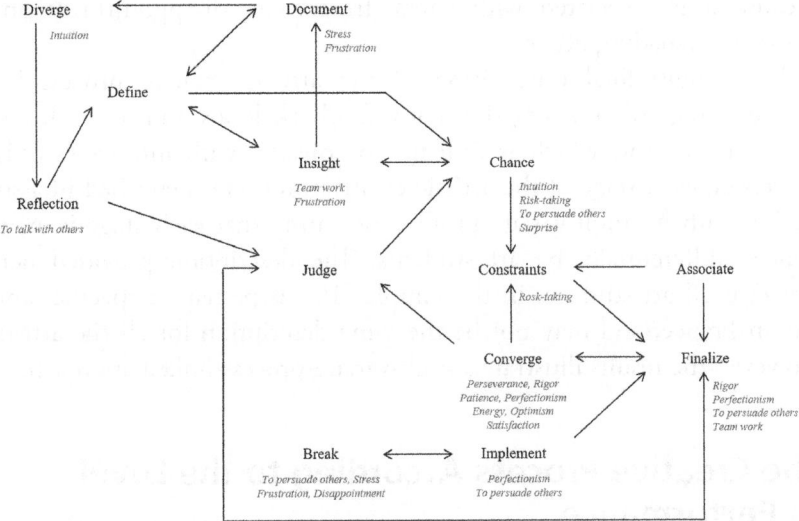

Fig. 3.1 Representation of the global artistic creative process and the multivariate factors involved in each stage (in step 2)

This stage is frustrating and tends to involve teamwork. *Associative thinking* leads to considering constraints and to finalizing the production. The *divergent thinking* stage is followed by reflection and is associated with intuition. Art students continue from *convergent thinking* to the consideration of constraints, judgement, and finalization. This stage is characterized by many multivariate factors (perseverance, rigor, patience, perfectionism, energy, optimism, and satisfaction). The stage of *benefiting from chance* leads typically back to the consideration of the constraints. It is associated with intuition, risk-taking, persuading others, and surprise. The *implementation* stage leads to finalization and breaks. This stage involves perfectionism and persuading others. *Finalization* is followed typically by convergent thinking and judgement, and it is associated with rigor, perfectionism, persuading others, and teamwork. The transition from *judgement* is common to "benefit from chance" and "finalization." Finally, *taking a break* is followed by implementation. This stage is experienced negatively by art students because it is associated with stress, frustration, disappointment and involves persuading others.

This longitudinal observation of the artistic creative process has the advantage of showing the many feedback loops and a "back and forth" movement, which is difficult to consider with interviews only. Moreover, each stage of the artistic creative process is described in association with it's multivariate profile, indicating that each stage is experienced differentially by art students. The description provided here concerns all art students in the sample. This is perhaps a specific sample, in France, and may not be the same description for all the artists. However, the results illustrate a multivariate-process-linked approach.

The Creative Process According to the Level of Performance

In addition to the global model treating all the art students together, there is some variability based on the level of performance. Indeed, using creativity ratings by the art teacher, it is possible to contrast two groups: art students with higher scores and art students with lower scores. Examining the proceedings of the creative process and the multivariate factor of each group, some variability appears (see Tables 3.1 and 3.2).

After defining the problem, the most creative art students go typically to document their ideas, whereas the less creative students start immediately with divergent thinking and have "insights." Even after documentation, the less creative group diverges, whereas the more creative students go back to the definition of the problem. Moreover, the less creative group starts and finishes by an insight, because transitions to this stage appear also after realization and finalization. These art students seem to have ideas too soon or too late, whereas the most creative students have insight after the convergent thinking stage. Additionally, the breaks are more efficient in this group because art students transit to associative thinking whereas the less creative students go to the stages of reflection or judgement of their work.

Table 3.1 Comparison of the transitions between the stages of the creative process according to the level of creative performance (in step 2)

Initial stage	Typically following stage(s)	
	Higher creative group	Lower creative group
Definition	Documentation	Constraints, insight, divergence
Reflection	Definition, implementation, **finalization**	Convergence, **finalization**, break
Documentation	**Definition**	**Definition**, divergence, convergence
Consideration of constraints	Judgment	Finalization
Insight	Chance, **break**	Judgment, **break**
Associative thinking	Judgment, convergence, finalization	Constraints, chance
Divergent thinking	Reflection, chance	Definition, documentation
Convergent thinking	Insight, association, divergence, chance	Constraints
Benefit from chance	Constraints, **divergence**	Definition, **divergence**
Implementation	Break	Constraints, insight
Finalization	**Convergence**	Insight, **convergence**, break
Judgment	Finalization	Chance
Taking a break	Association	Reflection, judgment

Note In bold are indicated the common transitions for both groups

Concerning the multivariate factors, the profiles are also very different between groups. Many stages are linked to intuition and deception in the less creative group (definition, reflection documentation, insight, realization, and judgement) whereas no stage is associated with intuition in the most creative group. Only the consideration of the constraints and the realization are disappointing for them. The less creative students seem blocked when they take a break because they need to ask for an opinion, whereas no multivariate profile is associated to this stage in the most creative group.

Globally, these variations of the artistic creative process and the multivariate factors involved according to the creative performance show that the most creative students have a more rational process, associated with stress at the beginning and deception at the end.

Table 3.2 Comparison of the multivariate profile associated to each stage of the creative process according to the level of creative performance (in step 2)

Stage	Main multivariate factors associated	
	Higher creative group	Lower creative group
Definition	Openness, to talk with others, team work	To ask for an opinion, to express ideas, to persuade others, disappointment
Reflection	To express ideas	Perfectionism, to talk with others, disappointment
Documentation	Patience, openness, satisfaction	Intuition, surprise
Consideration of constraints	To persuade others, stress, disappointment	Perseverance, energy, tolerance for ambiguity, frustration
Insight		Intuition
Associative thinking	Stress	Openness, intuition, risk-taking, surprise
Divergent thinking	Optimism	Patience, stress, frustration
Convergent thinking	To ask for an opinion, to express ideas	Perseverance, rigor, patience, perfectionism, energy, optimism, openness, risk-taking, satisfaction
Benefit from chance	Risk-taking	To express ideas, to persuade others
Implementation	Frustration, disappointment	Intuition, risk-taking, satisfaction
Finalization	Incertitude, **to persuade others, team work**	To talk with others, to express ideas, **to persuade others, team work**
Judgment	**Stress**, frustration	To talk with others, to ask for an opinion, to persuade others, **stress**, disappointment
Taking a break		To ask for an opinion

Note In bold are indicated the common multivariate profile for both groups

Maybe these students feel more motivated and are disappointed at the end not to have managed to realize their ideas, perhaps due to a lack of technical knowledge. In contrast, the creative process of the less creative art students appears more intuitive, associated with disappointment at the beginning and satisfaction at the end. Maybe these students are just happy to have done their work, even if the quality is not high enough.

Exercises to Improve the Artistic Creative Process

Based on interviews with professional artists and observations of art students in real contexts, we choose two factors to develop exercises improving and enhancing the creative process of art students: flexibility and social interactions (Botella & Lubart, 2018).

Flexibility was chosen because it is a main cognitive factor, essential in creativity (De Dreu, Bass, & Nijstad, 2007) and it was possible to develop exercises to improve it. Flexibility is the capacity to analyze a problem from several angles (Scott, 1999; Thurston & Runco, 1999). Lubart et al. (2015) stipulate that the flexibility is "the capacity to consider a single object, a single idea, under different angles, the sensibility to change as well as the capacity to move beyond an initial idea to explore new paths" (p. 37).

The second factor, social interactions, was chosen because in interviews, professional artists considered it as the least important for their creativity whereas observations of art students showed that social factors were frequently associated with stages of their creative process: they talked to others during the reflection stage, they worked in teams during insights, they persuaded others to benefit from chance and when they took a break, and they persuaded others and worked in teams during finalization. So, either the professional artists may have minimized the involvement of the social interactions in their process, or social interactions are only important for art students and not for professional artists. Although several studies highlighted the solitary (Cropley, 1999; Dogan, 1999; Storr, 1989), introverted, and individualist characters of artists (Feist, 1999), Csikszentmihalyi (2006) suggested that the social world is involved in the creativity. The artists are actually in touch with gallery owners and public. So, even if the artists can by shy and reserved, Feist (1999) underlines that nobody can be totally isolated.

Twenty-five art students participated in the study reported here (69% are females). First, the capacities of flexibility and social interactions of the students in art were estimated. On the basis of their results, the students were distributed in two groups: if they had more difficulties with flexible thought, they were assigned to the "flexibility" group ($n = 11$), whereas if they had relatively more relational difficulties, they were assigned to the "social" group ($n = 14$). Second, we proposed two exercises. So, students in the flexibility group had exercises allowing them to develop their flexible thought (they were invited to use various techniques, ones' never used or to try something new) whereas the students in social group had exercises during which they engaged in interactions with other students of the class (they were invited to discuss their ideas and productions with other students).

Art students completed a booklet (similar, but not identical, to the one described earlier). Nine stages of the artistic process were assessed: diverge, document, repeat, transform, destroy, associate, mark (engrave), finalize, and wait; and nine factors were included: logic, intuition, daydreaming, perseverance, divestment, openness, enjoyment, without hesitation, and sadness. Art students completed the booklet at each of the twelve studio sessions.

The Creative Process in the Different Exercise Groups

The artistic creative process of these two groups of students was observed to ascertain the most characteristic factors of every stage of the process. In Fig. 3.2, the most characteristic transitions between the stages (to which stage leads most frequently another stage) and the profiles of every stage (factors associated to every activity) are presented according to the group. These results underline that the creative process is different according to the group. However, they allow common points for the creative process of both groups to be highlighted. The divergent

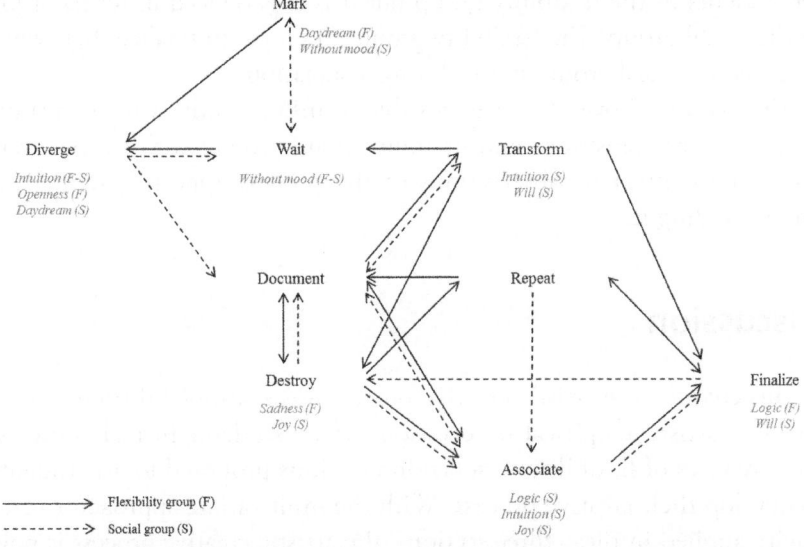

Fig. 3.2 Representation of the artistic creative process and the multivariate factors involved in each stage for flexibility and social groups (in step 3)

thinking stage involves intuition for both groups; the documentation is always followed by transforming or associating; the stage of destroying is always followed by documentation and association; after association, all students, regardless of the group, indicate documentation and finalization.

Except these common points, all the others transitions and multivariate profiles are specific to the group. When the waiting stage leads the flexibility group to diverge, it is the opposite transition for the social group. To diverge is typically followed by documentation for the social group, whereas this stage does not lead to a specific stage in flexibility group. The finalization leads the social group to the destructive stage whereas it interacts with repeating in the flexibility group. Additionally to the variety of transitions, the multivariate profile appears to differ also according to the group. Flexibility group daydreams during the marking (engraving) stage and the social group daydreams during the divergent thinking stage. The stage of destroying is related to sadness for art students in the flexibility group but it is experienced in terms of joy in the social group. The flexibility group uses logic to finalize their work whereas the social group uses it during association.

This section shows that it is possible to influence the artistic creative process (in its transitions and its multivariate profile) with simple exercises. The context of observations of the process appears essential for understanding it.

Discussion

In this chapter, the artistic creative process was examined through interviews of artists, empirical observations of art students in real contexts, and exercises of flexibility or social interactions proposed to art students to develop their creative process. With the multivariate approach to creativity applied in these three sections, the artistic creative process is now more understandable.

The Artistic Creative Process

From artists' discourse, we observe the stages of the creative process conceptualized as activity. The first stage of creative ideation is close to the problem-finding activities described by Getzels and Csikszentmihalyi (1976). The next stages of documentation/reflection, first sketches, and testing forms can also be found in various models (John-Steiner, 1997; Mace & Ward, 2002). More specifically, if a link is made with the Geneplore model (Finke, Ward, & Smith, 1992), the documentation/reflection phase corresponds largely to the generative process associated with the construction. In the same comparison, the search for ideas and first sketches and testing forms correspond to the explorative process—testing the creative idea. Finally, the testing forms stage was already described by Mace and Ward (2002) who explained that the artistic process is cyclical and iterative in nature, allowing the development of a series of products.

The observations of the artistic creative process in real contexts reveal that the subprocesses involved in creative work are not sequential, confirming the dynamic nature of the process: it is possible to skip a stage, to move back to a "previous stage," to realize several stages at the same time (Botella et al., 2011a). Graphic art students begin influenced by chance events. This specificity is supported by the model of Mace and Ward (2002) for whom artistic work begins by more or less vague impressions. According to our findings, the impression guiding the artistic process could be discovered by chance. Ayer (2010) also underlined the role of chance, which corresponds to chaos, freedom, and fantasy. Then, as found by Mace and Ward (2002), art students developed their idea, switching between divergent thinking, definition, documentation, reflection, insight, judgement, and chance. Next, art students implement their art work and finalize it. At this point, it is important to note the cyclic characteristic of the artistic creative process. Finalization can lead back to the judgement stage, confirming the cyclical and iterative nature of the artistic creative process.

Exercises proposed in the last section confirmed the results of the literature on the importance of cognitive flexibility for creativity (Botella et al., 2013; De Dreu et al., 2007; Levy & Langer, 1999). Even if results were not clear in the auto-questionnaire, the discourse of professional artists during interviews on their artistic approach underlined the beneficial influence of others. Although the artists are more individualistic (Feist, 1998, 1999), eccentric (Dogan, 1999), nonconformist (Barron, 1969; Barron & Harrington, 1981; Cropley, 1999; Gough, 1979; MacKinnon, 1965; McCrae & Costa, 1987; Roe, 1972) and "colder" than the general population (Getzels & Csikszentmihalyi, 1976), the social factors appear essential at least for art students because they are developing their personal and/or professional circle during their training. This result is supported by Bilton (2007) who suggested that the creative process is not only individual but also collective. In the present chapter, professional artists declared that others are important for their process (even if they are less important than cognitive factors) and it was possible to improve the social communication between art students.

The Multivariate Approach in Art

Spontaneously, during the interviews, artists make references to the key elements of the multivariate approach (cognitive, conative, emotional, and environmental dimensions). In addition to the discourse of artists, the analyses of questionnaires confirm the importance of the cognitive dimension for artistic creativity. The capacities of the individual to generate, to combine, to assemble, and to accept new ideas are all essential for creative production. The literature is consensual on the involvement of these cognitive capacities in creativity and the artists seem to agree as well. For example, Carlson and Gorman (1992) consider that selective comparison—that is the capacity to compare information and take into account previous knowledge—is very important for creativity. Bonnardel, Didierjean and Marmèche (2003) underline the importance of analogies and selective encoding in creative work. For artists in our sample, these cognitive abilities are also perceived as important.

The discourse of artists indicates that conative aspects are also favorable for creativity. The analysis of questionnaires confirms the fact that openness to new experiences, aesthetics, values, and sources of inspiration are important. In agreement with this, Chamorro-Premuzic, Furnham and Reimers (2007) explain that "personality differences underlying individuals' art interests seem to be captured mostly by the Openness to Experience dimension" (p. 85).

The impact of emotional factors in art seems to be a subject of controversy. During the interviews, artists mention often their emotions and how they could be helpful for their work. Nevertheless, in the questionnaires, the emotional dimension appears as the least important. Certain artists consider that emotional aspects are important for their creativity and others hold an opposite point of view. For example, some artists declare that satisfaction or anger could be the starting point for their creativity, whereas guilt is rather blocking. From the interviews with artists, it seems that both positive and negative emotions can facilitate creative work. However, when artists evoke the inhibitors of their creativity, emotional aspects are mainly cited. It can be concluded therefore that the importance of emotions and their favorable or unfavorable effects depends on emotional valence and specific content. This hypothesis can be completed by also considering the role of emotional intensity and emotional control.

The questionnaire results do not suggest that a particular environment is necessary for artists. Life events and finances are not seen as essential. Nevertheless, the social environment is indicated as the most important of all environmental dimensions. In artists' narratives, the social environment corresponds to the "social world," already described by the vertical and horizontal networks of Bilton (2013).

Additionally, observations of the multivariate factors in the real context of creativity show that factors are differentially associated to each stage of the process. The only common result to both observations of the creative process of art students and during the exercises is that intuition is involved in the divergent thinking stage. Moreover, results are consistent with the previous findings (Botella et al., 2011a). The first stages of the artistic creative process are difficult for students until they find an idea. They discuss with others to help their reflection. The stage of documentation is stressful and frustrating because, during their research, they see artworks of professional artists finished whereas they do not know what to do. In art, the technical constraints involve risk-taking because art students have to manage new materials. When, finally, the art students converge to a solution, they start to be satisfied and optimistic. However, it is important to note that it is not easy to converge, involving perseverance, rigor, patience, perfectionism, and tolerance for ambiguity. This is the stage with the most multivariate factors associated. Finally, the end of the artistic creative process involves persuading others of the chosen idea and perfectionism to realize and finalize it.

Finally, when we influenced the creative process by exercises, the flexibility group showed more openness with divergent thinking whereas the social group showed more daydreaming. Surprisingly, the stage of destroying and abandoning ideas involves sadness in the flexibility group whereas it involves joy in social group. Maybe social support could help students to accept that an idea is not good and had to be abandoned. Classmates could argue why an idea had to be removed whereas students had to make this reflection themselves in flexibility group, explaining why this stage is more difficult for these students compared to those in the social group.

Conclusion

The present chapter, based on in-depth analysis of the discourse of experienced artists, observations of art students and the proposition of exercises to help them to be more creative, offers a comprehensive exploration of creativity in art by triangulating several types of analysis. Whereas some of these analyses are common to creativity studies (e.g., the multivariate approach and discourse analysis), others suggest new theoretical perspectives (analysis of transitions and multivariate profile associated with each stage). These methods could be applied as well to other creative fields or specific art domains. Indeed, Fürst and collaborators (2012) showed visual art domain differences in the creative process among photographs, decorators, illustrators and designers. Moreover, the artistic experience could also be a line of investigation in the further research. Artists may not have the same creative process compared to art students as novices and experts create differently (Glăveanu, 2013). Fürst and collaborators (2012) observed that younger students started by illumination whereas advanced students started by preparation. The effect of the observational method could also be examined in future studies as Groenendijk, Janssen, Rijlaarsdam and van den Bergh (2013) have found a positive effect of experimenter's observations on the creative products, creative process, and motivation in visual arts. Based on such results, we can hypothesize that auto-observations could also have a positive effect on the artistic creative process.

Additionally, the findings presented offer new perspectives for the study and enhancement of creativity. It is thus hoped that a better understanding of the creative process and the factors involved could be helpful for teaching creativity and art. If some factors have unfavorable consequences for creative activity (as some emotional factors), it may be possible to control their effects, through training as experimented with flexibility and social interaction groups. The study of Stanko-Kaczmarek (2012) suggested also that providing exercises enhances positive affect, self-efficacy, and self-esteem. Moreover, if the creative process is clarified, ways of teaching more effectively creativity in art can be envisioned for each stage of artistic activity. These are certainly aspects that require further study and could very well use the findings of the present chapter as a starting point.

References

Amabile, T. M. (1983). Social psychology of creativity: A componential conceptualization. *Journal of Personality and Social Psychology, 45*(2), 357–377.

Amabile, T. M. (1996). *Creativity in context.* Colorado: Westview Press.

Arden, R., Chavez, R. S., Grazioplene, R., & Jung, R. E. (2010). Neuroimaging creativity: A psychometric view. *Behavioural Brain Research, 214*(2), 143–156.

Ayer, F. W. (2010). *The creative process in the individual—Which comes first, the egg or the hen?* World Congress on Communication and Arts, April 18–21, Guimarães, Portugal.

Baas, M., De Dreu, C. K. W., & Nijstad, B. A. (2008). A meta-analysis of 25 years of mood-creativity research: Hedonic tone, activation, or regulatory focus? *Psychological Bulletin, 134,* 779–806. https://doi.org/10.1037/a0012815.

Barron, F. X. (1969). *Creative person and creative process.* New York: Holt, Rinehart and Winston.

Barron, F., & Harrington, D. M. (1981). Creativity, intelligence, and personality. *Annual Review of Psychology, 32,* 439–476. https://doi.org/10.1146/annurev.ps.32.020181.002255.

Batey, M., & Furnham, A. (2006). Creativity, intelligence and personality: A critical review of the scattered literature. *Genetic, Social, and General Psychology Monographs, 132*(4), 355–429. https://doi.org/10.3200/MONO.132.4.355-430.

Bilton, C. (2007). *Management and creativity: From creative industries to creative management.* Oxford: Blackwells.

Bilton, C. (2013). Playing to the gallery: Myth, method and complexity in the creative process. In T. Kerry & J. Chan (Eds.), *Handbook of research on creativity* (pp. 125–137). Cheltenham: Edward Elgar.

Bonnardel, N. (1999). L'évaluation réflexive dans la dynamique de l'activité du concepteur [Assessing the dynamics of reflexive activity designer].

In J. Perrin (Ed.), *Pilotage et évaluation des activités de conception* [Piloting and evaluation of design] (pp. 87–105). Paris: L'Harmattan.

Bonnardel, N., Didierjean, A., & Marmèche, E. (2003). Analogie et résolution de problèmes. In C. Tijus (Ed.), *Métaphores et Analogies* (pp. 115–149). Paris: Hermès.

Botella, M., & Lubart, T. (2015). Creative processes: Art, design and science. In S. Agnoli & G. E. Corazza (Eds.), *Multidisciplinary contributions to the science of creative thinking* (pp. 53–66). New York: Springer.

Botella, M., & Lubart, T. (2018). Une recherche écologique pour développer la flexibilité et les relations sociales dans la créativité artistique. *Revue Française de Pédagogie, 197*, 13–22.

Botella, M., Zenasni, F., & Lubart, T. I. (2011a). A dynamic and ecological approach to the artistic creative process in arts students: An empirical contribution. *Empirical Studies of the Arts, 29*(1), 17–38. https://doi.org/10.2190/EM.29.1.b.

Botella, M., Zenasni, F., & Lubart, T. I. (2011b). Alexithymia and affective intensity of art students. *Psychology of Aesthetics, Creativity, and the Arts, 5*(3), 251–257. https://doi.org/10.1037/a0022311.

Botella, M., Glăveanu, V. P., Zenasni, F., Storme, M., Myszkowski, N., Wolff, M., & Lubart, T. (2013). How artists create: Creative process and multivariate factors. *Learning and Individual Differences, 26*, 161–170. https://doi.org/10.1016/j.lindif.2013.02.008.

Botella, M., Zenasni, F., & Lubart, T. (2015). Alexithymia and affective intensity of fine artists. *Journal of Creative Behavior, 49*(1), 1–12. https://doi.org/10.1002/jocb.54.

Botella, M., Nelson, J., & Zenasni, F. (2016). Les macro et micro processus créatifs. In I. Capron-Puozzo (Ed.), *Créativité et apprentissage* (pp. 33–46). Louvain-la-Neuve: De Boeck.

Botella, M., Zenasni, F., Nelson, J., & Lubart, T. (Forthcoming). Creative processes in five domains: Art, design, scriptwriting, music and engineering. In Lubart, T. (Ed.), *Homo creativus: The 7C's of creativity*. Singapore: Springer.

Busse, T. V., & Mansfield, R. S. (1980). Theories of the creative process: A review and a perspective. *Journal of Creative Behavior, 14*(2), 91–103. https://doi.org/10.1002/j.2162-6057.1980.tb00232.x.

Carlson, W. B., & Gorman, M. E. (1992). A cognitive framework to understand technological creativity: Bell, Edison, and the telephone. In J. J. Weber & D. N. Perkins (Eds.), *Inventive minds: Creativity in technology* (pp. 48–79). New York: Oxford University Press.

Carmeli, A., McKay, A. S., & Kaufman, J. C. (2014). Emotional intelligence and creativity: The mediating role of generosity and vigor. *The Journal of Creative Behavior, 48*(4), 290–309. https://doi.org/10.1002/jocb.53.

Chamorro-Premuzic, T., Furnham, A., & Reimers, S. (2007). The artistic personality. *The Psychologist, 20*(2), 84–87.

Cropley, A. J. (1999). Definitions of creativity. In M. A. Runco & S. R. Pritzker (Eds.), *Encyclopaedia of creativity* (Vol. 1, pp. 511–524). New York: Academic Press.

Csikszentmihalyi, M. (2006). *La créativité: psychologie de la découverte et de l'invention.* Paris: Robert Laffont.

De Dreu, C. K. W., Baas, M., & Nijstad, B. A. (2007, January). *Mood–creativity link revisited: Hedonic tone and activation level in the mood–creativity link.* Paper presented at the annual conference of the Society for Personality and Social Psychology, Memphis, TN.

Dietrich, A., & Kanso, R. (2010). A review of EEG, ERP, and neuroimaging studies of creativity and insight. *Psychological Bulletin, 136*(5), 822–848. https://doi.org/10.1037/a0019749.

Dogan, M. (1999). Marginality. In M. A. Runco & S. R. Pritzker (Eds.), *Encyclopaedia of creativity* (Vol. 2, pp. 179–184). New York: Academic Press.

Feist, G. J. (1994). The affective consequences of artistic and scientific problem solving. *Cognition and Emotion, 8*(6), 489–502. https://doi.org/10.1080/02699939408408955.

Feist, G. (1998). A meta-analysis of personality in scientific and artistic creativity. *Personality and Social Psychology Review, 2*(4), 290–309.

Feist, G. J. (1999). The influence of personality on artistic and scientific creativity. In R. J. Sternberg (Ed.), *Handbook of creativity* (pp. 273–296). Cambridge: Cambridge University Press.

Finke, R. A., Ward, T. B., & Smith, S. S. (1992). *Creative cognition: Theory, research, and applications.* Cambridge, MA: MIT Press.

Furnham, A., & Bachtiar, V. (2008). Personality and intelligence as predictors of creativity. *Personality and Individual Differences, 45*(7), 613–617. https://doi.org/10.1016/j.paid.2008.06.023.

Furnham, A., Batey, M., Booth, T. W., Patel, V., & Lozinskaya, D. (2011). Individual difference predictors of creativity in art and science students. *Thinking Skills and Creativity, 6,* 114–121. https://doi.org/10.1016/j.tsc.2011.01.006.

Fürst, G., Ghisletta, P., & Lubart, T. (2012). The creative process in visual art: A longitudinal multivariate study. *Creativity Research Journal, 24*(4), 283–295. https://doi.org/10.1080/10400419.2012.729999.

Gardner, H. (1993). *Creating minds: An anatomy of creativity seen through the lives of Freud, Einstein, Picasso, Stravinsky, Eliot, Graham, and Gandhi.* New York: Basic Books.

George, J. M., & Zhou, J. (2002). Understanding when bad moods foster creativity and good ones don't: The role of context and clarity of feelings. *Journal of Applied Psychology, 87,* 687–697. https://doi.org/10.1037/0021-9010.87.4.687.

Getz, I. W., & Lubart, T. I. (2000). An emotional-experiential perspective on creative symbolic-metaphorical processes. *Consciousness and Emotion, 1*(2), 89–118.

Getzels, J. W., & Csikszentmihalyi, M. (1976). *The creative vision: A longitudinal study of problem finding in art.* New York: Wiley.

Glăveanu, V. P. (2013). Creativity and folk art: A study of creative action in traditional craft. *Psychology of Aesthetics, Creativity, and the Arts, 7*(2), 140. https://doi.org/10.1037/a0029318.

Glăveanu, V. P., Lubart, T., Bonnardel, N., Botella, M., De Biasi, P.-M., De Sainte Catherine, M., ... Zenasni, F. (2013). Creativity as action: Findings from five creative domains. *Frontiers in Educational Psychology, 4.* https://doi.org/10.3389/fpsyg.2013.00176.

Goor, A., & Sommerfeld, R. E. (1975). A comparison of problem-solving processes of creative students and noncreative students. *Journal of Educational Psychology, 67*(4), 495–505. https://doi.org/10.1037/h0077009.

Gough, H. G. (1979). A creative personality scale for the adjective check list. *Journal of Personality and Social Psychology, 37,* 1398–1405. https://doi.org/10.1037/0022-3514.37.8.1398.

Groenendijk, T., Janssen, T., Rijlaarsdam, G., & van den Bergh, H. (2013). The effect of observational learning on students' performance, processes, and motivation in two creative domains. *British Journal of Educational Psychology, 83*(1), 3–28. https://doi.org/10.1111/j.2044-8279.2011.02052.x.

Gruber, H. E. (1981). *Darwin on man: A psychological study of scientific creativity* (2nd ed.). Chicago: University of Chicago Press.

Helson, R. (1999). Personality. In M. A. Runco & S. R. Pritzker (Eds.), *Encyclopaedia of creativity* (Vol. 2, pp. 361–371). New York: Academic Press.

John-Steiner, V. (1997). *Notebooks of the mind: Explorations of thinking* (Rev ed.). New York: Oxford University Press.

Lefebvre, L., Reader, S. M., & Sol, D. (2013). Innovating innovation rate and its relationship with brains, ecology and general intelligence. *Brain, Behavior and Evolution, 81,* 143–145. https://doi.org/10.1159/000348485.

Levy, B., & Langer, E. (1999). Aging. In M. A. Runco & S. R. Pritzker (Eds.), *Encyclopaedia of creativity* (Vol. 1, pp. 45–52). New York: Academic Press.

Lubart, T. I. (1999). Componential models. In M. A. Runco & S. R. Pritzker (Eds.), *Encyclopaedia of creativity* (Vol. 1, pp. 295–300). New York: Academic Press.

Lubart, T. I., Mouchiroud, C., Tordjman, S., & Zenasni, F. (2015). *Psychologie de la créativité* [Psychology of creativity] (2nd ed.). Paris: Armand Colin.

Mace, M.-A., & Ward, T. (2002). Modeling the creative process: A grounded theory analysis of creativity in the domain of art making. *Creativity Research Journal, 14*(2), 179–192. https://doi.org/10.1207/S15326934CRJ1402_5.

MacKinnon, D. W. (1965). Personality correlates of creativity. In M. J. Aschner & C. E. Bish (Eds.), *Productive thinking in education* (pp. 159–171). Washington, DC: National Education Association.

McCrae, R. R., & Costa, P. T. (1987). Validation of the five-factor model of personality across instruments and observers. *Journal of Personality and Social Psychology, 52*(1), 81–90. https://doi.org/10.1037/0022-3514.52.1.81.

Nakamura, J., & Csikszentmihalyi, M. (2003). The motivational sources of creativity as viewed from the paradigm of positive psychology. In L. G. Aspinwall & U. M. Staudinger (Eds.), *A psychology of human strengths: Fundamental questions and future directions for a positive psychology* (pp. 257–269). Washington, DC: American Psychological Association.

Osborn, A. F. (1963). *Applied imagination* (3rd ed.). New York: Scribners.

Patrick, C. (1935). Creative thought in poets. *Archives of Psychology, 178,* 1–74.

Patrick, C. (1937). Creative thought in artists. *Journal of Psychology, 4,* 35–73.

Roe, A. (1972). Patterns in productivity of scientists. *Science, 176,* 940–941.

Schlewitt-Haynes, L. D., Earthman, M. S., & Burns, B. (2002). Seeing the world differently: An analysis of descriptions of visual experiences provided by visual artists and nonartists. *Creativity Research Journal, 14,* 361–372. https://doi.org/10.1207/S15326934CRJ1434_7.

Scott, T. E. (1999). Knowledge. In M. A. Runco & S. R. Pritzker (Eds.), *Encyclopaedia of creativity* (Vol. 2, pp. 119–129). New York: Academic Press.

Shaw, M. P., & Runco, M. A. (1994). *Creativity and affect*. Westport: Ablex Publishing.

Stanko-Kaczmarek, M. (2012). The effect of intrinsic motivation on the affect and evaluation of the creative process among fine arts students. *Creativity Research Journal, 24*(4), 304–310. https://doi.org/10.1080/10400419.2012.730003.

Sternberg, R. J., & Lubart, T. I. (1991). An investment theory of creativity and its development. *Human Development, 34,* 1–31. https://doi.org/10.1159/000277029.

Sternberg, R. J., & Lubart, T. I. (1995). *Defying the crowd: Cultivating creativity in a culture of conformity*. New York: Free Press.

Sternberg, R. J., & O'hara, L. A. (2000). Intelligence and creativity. In R. J. Sternberg (Ed.), *Handbook of intelligence* (pp. 252–272), New York: Cambridge University Press.

Storme, M., Myszkowski, N., Çelik, P., & Lubart, T. (2014). Learning to judge creativity: The underlying mechanisms in creativity training for non-expert judges. *Learning and Individual Differences, 32,* 19–25. https://doi.org/10.1016/j.lindif.2014.03.002.

Storr, A. (1989). *Solitude*. New York: The Free Press.

Tegano, D. W. (1990). Relationship of tolerance of ambiguity and playfulness to creativity. *Psychological Reports, 66*(3, Pt 1), 1047–1056. https://doi.org/10.2466/pr0.1990.66.3.1047.

Thurston, B. J., & Runco, M. A. (1999). Flexibility. In M. A. Runco & S. R. Pritzker (Eds.), *Encyclopaedia of creativity* (Vol. 1, pp. 729–732). New York: Academic Press.

Treffinger, D. J. (1995). Creative problem solving: Overview and educational implications. *Educational Psychology Review, 7*(3), 301–312. https://doi.org/10.1007/BF02213375.

Wallas, G. (1926). *The art of thought*. New York: Harcourt-Brace.

Wolfradt, U., & Pretz, J. E. (2001). Individual differences in creativity: Personality, story writing, and hobbies. *European Journal of Personality, 15*(4), 297–310. https://doi.org/10.1002/per.409.

Wolfradt, U., Felfe, J., & Koster, T. (2002). Self-perceived emotional intelligence and creative. personality. *Imagination, Cognition and Personality, 21*(4), 293–310.

Yokochi, S., & Okada, T. (2005). Creative cognitive process of art making: A field study of a traditional Chinese ink painter. *Creative Research Journal, 17*(2&3), 241–255. https://doi.org/10.1080/10400419.2005.9651482.

Zenasni, F., & Lubart, T. I. (2001). Adaptation française d'une épreuve de tolérance à l'ambiguïté: Le M.A.T [French adaptation of a test of tolerance of ambiguity: The Measurement of Ambiguity Tolerance (MAT)]. *Revue Européenne de Psychologie Appliquée* [European Review of Applied Psychology], *51*(1–2), 3–12.

Zenasni, F., & Lubart, T. I. (2008). Emotion related-traits moderate the impact of emotional state on creative potential. *Journal of Individual Differences, 29*(3), 157–167. https://doi.org/10.1027/1614-0001.29.3.157.

Zenasni, F., Besançon, M., & Lubart, T. I. (2008). Creativity and tolerance of ambiguity: An empirical study. *Journal of Creative Behavior, 42*(1), 61–73. https://doi.org/10.1002/j.2162-6057.2008.tb01080.

4

The Creative Process in Writers

Jane Piirto

The creative writer is a kind of creator who has earned a certain fascination among people, especially readers, and especially avid readers. What does the writer do to create such interesting, entertaining, engaging, profound, scary, descriptive, prescient, insightful, even depressing thoughts put together into sentences that take form in fiction, nonfiction, poetry, film, song, and plays? How does the writer work? Why does the writer write—or not write? What happens before, during, and after writing—in other words, what is the creative process by which such artifacts emerge?

Creativity researchers, who are usually engaged in and who have studied in the academic discipline of psychology, are presented with an interesting dilemma—how to account for the miracle of the artistic written word with the often Latinate and jargon-filled language of the science of psychology? How can they explain the creative writer, who is often unwilling and may even be scornful of being studied?

J. Piirto (✉)
Ashland University, Ashland, OH, USA
e-mail: jpiirto@ashland.edu

© The Author(s) 2018
T. Lubart (ed.), *The Creative Process*, Palgrave Studies in Creativity and Culture,
https://doi.org/10.1057/978-1-137-50563-7_4

For example, the poet Kenneth Rexroth, who had been invited and who accepted an invitation to be studied along with other eminent writers at the Institute for Personality Assessment and Research wrote a bitter article about the experience, "The Vivisection of a Poet" (Barron, 1968). About a fifth of the writers studied objected; they thought psychological research was "intrinsically evil" and "presumptuous because it seeks to describe and to understand what is intrinsically a mystery" (Barron, 1995, p. 68). I have experienced similar reactions from literary writers. When I gave copies of my 2002 study to a group of fellow writers at a writers' conference, one responded to my gesture and to my accompanying lecture with a sardonic poem that she later published and addressed to me.

What is the creative process and why even bother about it? The creative process has engaged theorists and researchers about creativity since the beginnings of the field as an entity within the discipline of psychology. One of the earliest writers on creativity, Donald W. MacKinnon (1978), who led the Institute of Personality Assessment and Research (IPAR) in the mid-1950s at the University of California-Berkeley wrote, about the creative process, that it can be studied in many ways: "retrospective reports, observation of performance on a time-limited creative task (e.g., writing a poem), factor analysis of the components of creative thinking, experimental manipulation … of variables … relevant to creative thinking, simulation … on the high-speed electric computer" (p. 190). He thought that a study of introspective accounts by creators, hypnosis, a study of incubation or "unconscious processing," the role of chance, the role of "incidental cues" in solving problems creatively, an inquiry into the relative length of the various stages and phases within the length of the whole process, should be considered. He also thought that studying from a psychoanalytic, psychophysiological, and even a study of the influence of mind-altering drugs should be part of the repertoire of researchers.

MacKinnon (1978) called for research into the creative process in various domains (for which the present volume is an example), and he questioned "whether creativity is always a matter of problem-solving," as "one of the salient traits of a truly creative person is that he sees a problem where others don't" (p. 195). MacKinnon questioned the accuracy of saying that the creative process, especially of visual artists, composers, poets, and other artists begins with conceptualizing and formulating some kind of problem. He wondered whether Getzels and Csikszentmihalyi's (1976) study of problem-finding in painters was valid: "I'm not so sure" (p. 195), saying "The artist has a need to express himself, to resolve some tension or imbalance, to do something with the materials of his art. However, does he have a problem to solve, and if he does in what sense is it a problem?" MacKinnon's prescient evaluation of what studies need to be done about the creative process still holds.

Likewise, Frank Barron (1968), studied the creative writers at IPAR 1957–1961, describing the creative process in writers. In a long and multifaceted definition of what the creative process in writers consists, he wrote:

> A secondary purpose was to investigate the process of creation in writing through careful study of an author's work, through intensive interviews with him about his work, and through tests calling for composition, providing an opportunity for creative perception and expression. Process was taken to include these aspects of the act most prominently: the conscious intention of the writer; preconscious or unconscious intentions, and determinants of the conscious intentions (such as the psychic needs being served, the origins of fantasy, the meaning of the work in relation to the total life cycle of the writers); the choice of form; significant revisions, discarded beginnings, final self-criticism; unexpected or unexplained changes in intention or form, sudden inspiration; temporal and emotional phases in the process (intensity of work and feeling, blocks, distribution of attention, flow, alternation of convergent and divergent phases), feeling of completion or incompletion, attitude toward the work when it is completed. (p. 238)

Who are the writers studied in these studies? They are writers who, as Barron (1968) said, engage in "the composition of phrases, essays, stories, poems, and plays which communicate a single individual's interpretation of experience in an original manner" (p. 238). A more contemporary view would also include screenwriters and songwriters—for example, Bob Dylan, Leonard Cohen, William Goldman, and Aaron Sorkin. The IPAR work was one of the most complicated studies of the creative process in writers as it included a comparison group of writers who were not engaged at such a high level of eminence—all of these aspects of the creative process in writers did not seem to be discussed in the ensuing publications, but seemed to mostly focus on the results of the paper and pencil assessments of personality such as the Adjective Check List (ACL), California Psychological Inventory (CPI), and the Myers–Briggs Type Indicator (MBTI). In 1995, Barron listed some of the writers who were studied and who had died so he could release their names: William Carlos Williams, Marianne Moore, Arthur Koestler, Muriel Rukeyser, Jessamyn West, MacKinlay Kantor, Joseph Wood Krutch, A. B. Guthrie, Jr., and Truman Capote.

The following essay engages the creative process in creative writers with a little different stance than that which is usual for chapters such as this. The author hereby presents herself with an innate dual bias—she is an academic who writes about the psychology of creativity; but she has also published and is an award-winning creative writer of poetry, fiction, and nonfiction.

Current Theories of the Creative Process

This volume has ample explication of the creative process. Of special applicability for the present chapter, here is an overview of three sources: (1) the findings about the creative process in writers of the IPAR studies in the 1950s; (2) the summaries by Lubart (2009); and (3) in addition, the present author (Piirto, 2002) has written a chapter about the creative process in a book-length study of 160+ U.S. creative writers who met a high standard of publication.

Method

The method by which the following have been obtained is qualitative archival. The author has read many interviews, biographies, and memoirs of creative writers and has gleaned from them certain commonalities—called themes—about the creative process in writers. As Vera

John-Steiner (1997) said, "The method of self-reports, I have found, is an effective means of learning about scientific thinking" (p. 181). In my book, *"My Teeming Brain": Understanding Creative Writers* (Piirto, 2002), a qualitative study of 160 men and women U.S. writers who had met the publication requirements to be able to be listed in the annually updated *Directory of American Poets & Writers*, I had a special chapter for the creative process in creative writers, which is a predecessor of this chapter. It has been paraphrased and reorganized here, except for direct quotations from writers. The writers discussed in the present chapter also meet those criteria, which are these: You must have 12 points in a category to be listed; i.e., you must have 12 poetry credits to be listed as a poet.

The following count as points for listing:

1. Each book of poetry, fiction, or creative nonfiction (personal essays or memoirs) (12 points)
2. Each chapbook (6 points)
3. Each work of fiction or creative nonfiction (personal essays or memoirs) published in a literary journal, anthology, or edited Web publication (2 points)
4. Each spoken word performance (not readings) (2 points)
5. Each poem published in a literary journal, anthology, or edited Web publication (2 points) (http://www.pw.org/directory/criteria)

I myself meet these criteria in both poetry and in fiction and am listed in the *Directory*. Since the 2002 work, I have organized my scholarly work on the creative process into Five Core Attitudes about creativity; Seven I's for the creative process; and General Practices in the creative process (Piirto, 1998, 2002, 2004, 2008, 2009, 2010, 2011a).

Motivation

The will to write is first. None of the creative process exists without motivation. Why would a person decide to write poems, stories, novels, plays, songs, or films? The motives vary, but the writing life is hard and what begins as a need to communicate an idea or insight does not always continue. The rejection rate for submissions to literary journals is high, especially from those journals that are most prestigious. It is extremely hard to find a literary agent. The submission process takes an accountant-like sense of organization, for many journals do not accept simultaneous submissions, or if they do, they ask the writer to withdraw the submission if a piece is accepted elsewhere. For writers who have many manuscripts submitted at the same time, an organized system is necessary. Lucia Nevai, a fiction writer, had a chart where she followed the progress of at least ten short stories at a time through ten submissions to ten different literary journals (personal communication, Lucia Nevai, 1988). The passion, the motivation, is what I have called the "thorn" on my Piirto Pyramid of Talent Development (see Fig. 4.1), that which the creator can't not do. Cynthia Ozick said, "the beginning was almost physiological in its ecstatic pursuits" (Teicholz, 1988, p. 297). She said it was "delectable excitement," "waiting-to-be-born excitement. I suppose it was a kind of parallel Eros."

Five Core Attitudes for Creators

Self-Discipline

Writers who produce must be disciplined. They must put their lives into a schedule, and they must write as if they are going to work. Essayist E. B. White said,

> In the end, a man must sit down and get the words on paper and against great odds. This takes stamina and resolution. Having got them on paper, he must still have the discipline to discard them if they fail to measure up; he must view them with a jaundiced eye and do the whole thing over as many times as it necessary to achieve excellence or as close to excellence as he can get. (Plimpton & Crowther, 1988, p. 13)

Alice Munro, the Nobel Prize for Literature short story writer, described how disciplined she was while writing as a full-time mother and a

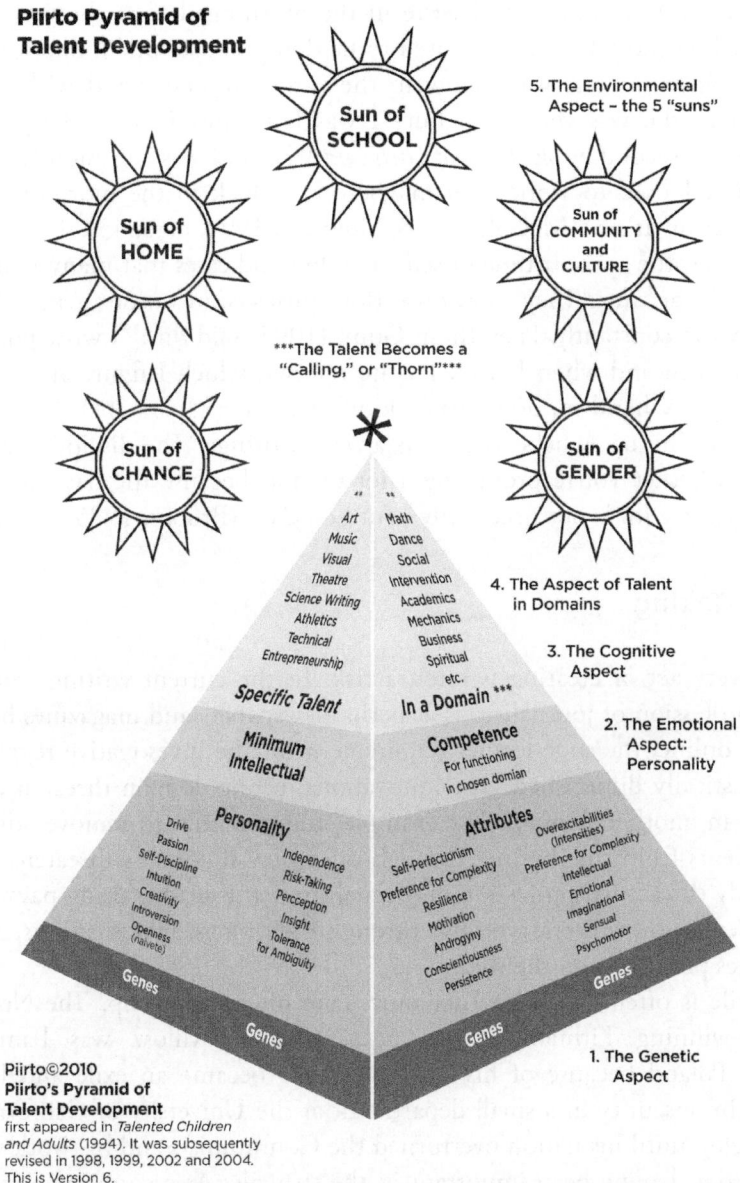

Fig. 4.1 Piirto Pyramid of Talent Development (*Source* Jane Piirto, chapter author, used with permission from the author)

family business. She would write in the mornings before she was due at the bookstore ("I was supposed to be doing housework"). She would write when no one was home, in the early afternoon, and with four children, she was very busy, but "I was fairly productive. The year I wrote my second book, *Lives of Girls and Women*, I was enormously productive. I used to work until maybe one o'clock in the morning and then get up at six" (McCullough & Simpson, 1994, para. 43).

In fact, the core attitude of self-discipline indicates that many writers write all the time. They work on various projects, in various genres, but they write constantly. Poet Thom Gunn (1995) said this: "I write poetry when I can and when I can't, I write reviews, which I figure at least is keeping my hand in, doing some kind of writing" (para. 11). Nigerian novelist Chinua Achebe said he is always writing: "The discipline is to work whether you're producing a lot or not, because the day you are producing a lot is not necessarily your best day" (Brooks, 1995, p. 158).

Risk-Taking

The very act of creating is often a risk. In the current writing world, the profession of journalism is at peril. Newspapers and magazines have gone online; the shoe leather requirement for the investigative reporter is drastically diminished. The innovations in one domain threaten creators in another domain. For example, the potential to remove advertisements from the internet through one's new iPhone 6s threatens the already threatened profession of journalism because advertising pays the salaries for the writers, who live through their blogs, investigations, and articles published on the web.

Exile is often a risk a writer must take for speaking up. The Nobel prize-winning Lithuanian-born poet, Czeslaw Milosz was banned from Poland because of his early work; he became an exile and labored in obscurity in a small department at the University of California, Berkeley, until his nation overturned the Communists and his work was viewed as having been important in the struggle. As a young man who had emigrated to Poland, he worked for the Polish embassy in Paris but then quit and became a refugee in Paris. His stance was risky in terms of his career, for the Paris writing establishment distrusted him: "I was on friendly terms with Albert Camus at the time Jean Paul Sartre and his crowd were after him, trying to destroy him because in *The Rebel* ... he had mentioned there were concentration camps in the Soviet Union" (Faggen, 1995, p. 260). Sartre and other writers in Paris were at this time avid Communists who did not believe Stalin was doing what he did in the gulags and concentration camps.

Writers are often forced to risk exile because of what they say. Nobel prize winner Wole Soyenka conjectured why, after exiled Bangladeshi writer Taslima Nasrin returned back home even though she was in danger. When she got there she was imprisoned. Soyenka said her risk-taking illustrated

> ... that periodic need of all creative people to re-cross the threshold of loss, to recharge the batteries of identity and thus engage in the ritual of the lifting of the creative frontier, an attempt yet again to wage war within the liminal zone, the writer's normal place of habitation that sometimes turns unbearably physical. Those creative frontiers ... remain "territories of hazardous navigation for the voyager, the writer, who remains a suspect émigré in a refugee camp and whose status of semi-exile undergoes quite arbitrary forms of articulation." (para. 11)

Another Nobel prize winner, Aleksandr Solzhenitsyn, was also imprisoned and exiled. His ground-breaking work, *One Day in the Life of Ivan Denisovich* (1963) describing the conditions in the Siberian work camps demonstrated the acute risk-taking that certain particular writers went through and what they endured.

What does this have to do with the creative process? It illustrates what Barron (1968), who was quoted above, called the "meaning of the work in the total life cycle of the author" (p. 238). For these and many other writers, the production of the work is worth taking risks of exile, imprisonment, or censure.

However, the trope that artists (writer) are risk-takers was demeaned by Woody Allen as being solipsistic. He said, "I find funny and silly the pompous kind of self-important talk about the artist that takes risks." He said that the real risk-takers among artists are "the people who took risks against the Nazis or some of the Russian poets who stood up against the state—those people are really courageous and brave" (Kakutani, 1995, p. 216). He said that such risk was real risk, not "casting against type, wow, what danger!" (p. 216). However, other writers who are not in exile and not in jail, may have what Woody Allen disparaged, what Rollo May called "creative courage," the courage to try new things, and to try to be in the vanguard, to be an early warning of what is coming into the culture, showing, through image and through symbols, what is about to happen in the world, using the imagination to predict what is possible. In literature, we can cite the innovative stream-of-consciousness style of James Joyce in *Ulysses*; the New Critics influence on the avant-garde such as the fiction of Donald Barthelme in *Sixty Stories* (1981); the postmodern form—amply footnoted—of David Foster Wallace in his *Infinite Jest* (1996); the *Reality Hunger* manifesto by David Shields (2010). In poetry, the French vers librists, who freed the poem from rigid line and opened up white space; the Imagists, led by Ezra Pound; the Modernists (such as T. S. Eliot); the Beat poets such as Allen Ginsberg; the innovative long-verse line work of Jorie Graham (1995); and the hip-hop vibe of poetry slams have all moved the field of creative writing forward through the taking of creative risks.

The subject matter itself might involve risk-taking. Poet Anne Sexton's confessional style of poetry exposed many of her family and personal secrets. Asked why she did this, she responded: "It took a certain courage, but as a writer one has to take the chance on being a fool... yes, to be a fool, that perhaps requires the greatest courage" (Kevles, 1989, p. 281).

Group Trust

One would think that the creative writer, solitary, in his or her writing room (a room of one's own, Virginia Woolf) would not need or be susceptible to needing and having trust among and from the group. The literature consulted for this study suggests otherwise: pods of writers who exchange work, who critique each other's work, who attend each other's performances and parties, who ask and give advice to each other exist.

One of the most gossipy biographies consulted, Hart's (2000) biography of James Dickey, positively jumped with name-dropping of poets and writers of the 1950s and 1960s with whom Dickey associated, and with whom he had trust and mistrust, depending on the day: David Ignatow, Robert Bly, John Berryman, Richard Hugo, James Wright, Donald Hall, Carolyn Kizer, Malcolm Cowley, Rosemary Daniell, and many others. These writers knew each others' work and discussed their theories and the ups and downs of publication. There was a thinkers' war between those who thought like the New Critics and those who thought poetry should take more archetypal and mythic consciousness. Friendships between writers take shape according to the hierarchical echelon they inhabit. They meet each other at readings and at workshops and socialize with each other, but not with lesser writers; for example, the hierarchical dining arrangements at the Bread Loaf Writers' Conference, where the students almost never dine with the famous teachers has led some writers to feel quite inferior (Judith Lindenau, personal communication, August 1977).

Tolerance for Ambiguity

One of the humanistic psychologists who studied creativity, Abraham Maslow, described how tolerance for ambiguity is related to the creative process: "creativeness is correlated with the ability to withstand the lack of structure, the lack of future, lack of predictability, of control, the tolerance for ambiguity, for planlessness" (Maslow, 1998, p. 239). The creative writer lives with ambiguity throughout the creative process. In my research, I have found very few (perhaps early Norman Mailer and best-seller Irving Wallace, who described his research and outlining for the novel, *The Prize* in (1962) in *The Writing of One Novel*, 1968) who meticulously plan their work and view it as a problem to be stated and thus solved. Most writers live in a state of ambiguity, especially in the early stages of their composition process. Some have stated they know the first sentence and the last sentence, but getting there is an ambiguous process. Novelist Mark Helprin is an example. He described his process:

I certainly know how it will end. In fact, I build everything toward the last sentence, which is the first thing that occurs to me in writing a book. It's like throwing a stone into a lake and then swimming and diving to fetch it. … Knowing the beginning and the end means that the middle is where the surprises are. (Linville, 1993, p. 195)

However, even disciplined screenwriter/playwright Aaron Sorkin, while writing weekly series such as *The West Wing* and *The Newsroom* have stated that they do not know how things will end and where they are going: "Lots of times, I start writing without knowing how it ends, and I call those 'Skylab' episodes … from time to time, I'll write a Skylab episode where I'll start not knowing how I'm going to finish, but just hoping that somewhere in the middle it's going to occur to me how this ends" (Daniels, 2012, para. 14). IPAR found that the writers, like other creators, "are born with greater brain capacity; they have more ability to hold many ideas at once, and to compare more ideas with one another—hence, to make a richer synthesis" (Barron, 1995, p. 67).

Openness to Experience

Intuition is related to Openness to Experience on the Big-Five NEO-PI-R (Langan-Fox & Shirley, 2003). Beyond such assessment evidence, writers continually talk about how they read and study and learn and try to understand others' points of view. They are also remarkably open to trying new things, especially in the world of the arts and culture. Writer Robert Love Taylor (1993) talked about going to San Francisco after a sheltered youth spent in Oklahoma and Utah.

In San Francisco I fell among literate friends who urged on me their books and their enthusiasms. Their enthusiasms were easy enough for me to embrace: existentialism, zen poetry, wine, low-priced ethnic cafes (Chinese, of course, and Italian, but also Basque and Filipino), flamenco guitars and Delta blues, Dostoevsky, foreign films (especially Fellini and Bergman), Kandinski and Klee, and of course the left-wing politics … (p. 263)

Seven "I's"

I have used the convenience of alliterative list-making to focus on some parts of the creative process that begin with "I." These are several types of inspiration, imagination, intuition, insight, imagery, improvisation, and incubation.

Inspiration

What does it mean to be inspired? Inspiration in the creative process is widely spoken of. In the sense of inspiration being a "thing" or a "kind" it is perhaps convenient and will qualitatively clarify the inspirational experience. I have categorized several types of inspiration that seem to be common with writers.

Inspiration from Love: Visitation of the Muse

That writers write to explain or respond to an attraction, overtly or covertly sexual, undoubtedly emotional, is not news. Perhaps the most famous, the *Sonnets from the Portuguese*, Elizabeth Barrett Browning's response to meeting and corresponding with Robert Browning— "How do I love thee? Let me count the ways" (Sonnet 14, 1850); or Shakespeare's "My love is as a fever, longing still/ For that which longer nurseth the disease": (Sonnet 116) will serve to illustrate the omnipresence of love and desire as an inspiration for writers.

The muse is the ideal lover: she [he] is perfect and has the wherewithal to promote obsession in the writer. I have written elsewhere at length on inspiration by the Muse (Piirto, 2002, 2004, 2011a). One of the most famous muses is Suzanne in the song of the same name, written by poet/singer-songwriter Leonard Cohen. In her biography of Cohen, Simmons (2012) noted that Suzanne, last name Verdal, was a well-known bohemian in Montreal, who inspired other artists as well. Cohen said

Everything happened just as it was put down. She was the wife of a man I knew. Her hospitality was immaculate. ... An old friend of mine whose name was Suzanne invited me down to her place near the river. ... The purity of the event was not compromised by any carnality. The song is almost a reportage. (Simmons, 2012, p. 129)

Cohen also wrote songs for other muses; songwriters Judy Collins (who first covered "Suzanne)"; songwriter Nico (who also inspired Bob Dylan, Jackson Browne, Jimi Hendrix, and Jim Morrison), and Joni Mitchell. Then Mitchell wrote for Cohen, and he became her muse. Simmons (2012) said, "For the first time the tables were turned: Leonard was the muse for a woman. Not just any woman but one whom David Crosby ... calls 'the greatest singer-songwriter of our generation'" (Simmons, 2012, p. 172).

Inspiration from Dreams

The act of writing has been called the waking dream by many writers; this has to do with the sense of being in a trance that envelops the writer while writing. For example, Stephen King said, "part of my function as a writer is to dream awake" (King, 1993, p. 145). Novelist Charles Johnson said, "writing is a lot like dreaming. You're creating a dream for the reader" (Johnson, 1993, p. 128). I will discuss this mystical/mysterious feeling later.

The more concrete inspiration of dreams is also important to writers. Often specific dreams provide material for writers. Poet Reynolds Price described how transcriptions of his dreams from his dream notebook landed intact, essentially, on the poetic page: he said that he has several poems in his books that are entitled "Dream Of ——"; "I'd sit down and write them up as straight-forward poems, not offering any of the details of the dream narrative" (Price, 1993, p. 202). However, these were poems, and Price couldn't "think of a time in which the idea for a particular novel or short story came to me completely in a dream" (p. 206). For Clive Barker, though, some of his dreams become "the starting place for a narrative and then you backtrack from that" (Barker, 1993, p. 34). Similarly, for Bharati Mukherjee, "In the stories I've ended up liking, and in the novel *Jasmine*, the endings have definitely come to me in dreams" (Mukherjee, 1993, p. 161). Novelist John Nichols also said something similar: "I've used dreams in different places in writing, almost word for word … I embellished the dream. I changed the context, changed the personality that was involved in it" (Nichols, 1993, pp. 179–180).

Inspiration of Novel Surroundings: Travel

Travel contains the lived experience of novelty, and with novelty, the observer is open to experience and experiences the world with a sense of naiveté. Picture such famous travel writers as Paul Theroux and Bruce Chatwin in the railroad cars they rattled around in while they traveled the world, the fields and hills and towns passing by, the stopping by accident or will. They scribbled in their notebooks and sent back dispatches to their publishers as they went. The novels of Patricia Highsmith detail her world travel as setting in intriguing plots.

Travel seems to facilitate the creative process of writers. The lure of the open road, lighting out for the territory, leaving the old hometown behind, are all themes in the literature. As a writer, I concur. This has something to do with a sense of freedom and something to do with imagination. Imagining place, or setting, is more difficult than imagining plot. Our visual culture thrives on images of far off places, and our nightly news shows us disasters in Tibet and tsunamis in Japan, old cars in Cuba, and the black flags of Isis flying above desert landscapes from vehicles made in the U.S. However, the physicality of being there cannot be duplicated in two-dimensional images.

A two-day visit to the northeast regions of Pakistan on the border of Afghanistan on Easter weekend in 1987 is what I rely on every time I see the news from the area. I even began a screenplay based in the region that had to do with a daughter of the Mujahadeen who was a teacher at the American School in Islamabad, while subsequently leading the rebellion against the Russians. I felt able to write about these far regions of the majestic Himalayas, as I had actually stood by the side of a road in upper Pakistan being clucked at with disgust by an old goatherd in a turban, next to a rushing aqua-limestone-water river, among the piles of pale brown rocks where nothing green could grow. Terrorist hideouts in 1998, I was easily able to picture the terrain targeted. When the earthquake destroyed the cathedral in Christchurch, New Zealand, I recalled attending a concert there by a piano prodigy and mourned the ruin of this famed and impressive landmark.

Inspiration from Others' Work

The famed *Paris Review* interviews with writers have as a common question, "Who has influenced you?" This would be the prewriting phase of the creative process, and could cover years before the writing process begins. The writers answer with enthusiasm and generosity. They tell stories and tales, gossip, praise, and malign each other. Very few writers do not read the work of their contemporaries, rivals, and predecessors. Among the most amusing was the frank British writer Rebecca West in her interview. Rebecca West was also a prolific reviewer of books and her tart judgment was much feared and praised. Here are excerpts:

- On Colette: "I didn't like her very much as a person and I think she was repetitive and I hate all her knowing nudges about men but I think she was a good writer on the whole." (Warner, 1989, p. 89)
- On Mark Twain: "I longed, when I was young, to write as well as Mark Twain. It's beautiful stuff." (p. 81)
- On T. S. Eliot: "Whom I didn't like a bit. He was a poseur." (p. 94)
- On Somerset Maugham: "He couldn't write for toffee, bless his heart." (p. 94–95)
- On Kafka: "Couldn't write about sex or value its place in life." (p. 95)
- On Lawrence: "An awful lot of nonsense … when he writes about Mexican sacrifices and sexual violence." (p. 95)
- On Ivy Compton-Burnett: "Had her own stereotype and wrote too many books exactly like each other in form. But it was a damn good form." (p. 98)

Another creative writer who reviews others and is influenced in her creative process by their work is Joyce Carol Oates, who listed both early influences and later ones. "I've been reading for so many years, and my influences must be so vast" (Phillips, 1989, p. 377). She listed Thoreau "whom I read at a very impressionable age (my early teens) and Henry James, Flannery O'Connor, Katherine Anne Porter, and Dostoevsky."

Edna O'Brien was also moved by the Russians: "from one page of Dostoevsky I feel renewed, however depressing the subject" (Guppy, 1989, p. 343). For her, Chekhov, though, topped Dostoevsky: "When I first read Chekhov's short stories before I saw his plays, I knew I had heard the voice I loved most in the whole world" (p. 344). She was puzzled by how his "dramatic genius was so mysterious; he does what seems to be impossible" (p. 344).

It might not be too hyperbolic to say that part of the prewriting process of almost all writers comes from their reading. For Robert Stone it was Kerouac's *On the Road*: Stone said it prefigured his own book, *A Hall of Mirrors*: "*On the Road* twenty years later. In a way. Yes" (Wood, 1988, p. 357). And one more example: For John Hersey, "Malraux, Silone, John Dos Passos … Hemingway, Faulkner were all writers who had excited me: the kind of skepticism and challenging of the norms that Van Santwood [a teacher who influenced him] had put to me had attracted me to writers who were trying to break the molds in various ways" (Dee, 1988, p. 112). Examples of how the work of other writers has influence on writers writing are legion.

Intuition

Intuition implies having a hunch. "Just knowing." A gut feeling. What does this have to do with creativity? Creative people trust their intuition. They prefer to use their intuition. Carl Jung theorized that people prefer either Intuition or Sensing. Writers scored high on Intuition (N) on the Myers–Briggs Type Indicator; 90% preferred Intuition as compared to 25% in the general population (Barron, 1968, 1995).

Everyone has intuition, but everyone does not trust their intuition. Intuition is indefinite, vague, whereas insight (which will be discussed below) is a brilliant flash. Policastro (1999) said that "creative intuition may be technically defined as a tacit form of knowledge that broadly constrains the creative search by setting its preliminary scope" (p. 92). Intuition guides but does not direct.

Poet Gary Snyder (1995) discussed "wild intuition" and imagination in an interview with Bill Moyers. He said, "I go to meet that blundering, clumsy, beautiful, shy world or poetic, archetypal, wild intuition that's not going to come out into the broad daylight of rational mind but wants to peek in" (p. 157). The importance of intuitive perception of the world, of a non-concrete but still tangible apprehension of underlying truth informs the creator's view of life. One is reminded of the Jungian psychological concept of synchronicity (see Piirto, 2011b), that all coincidences are not coincidental, but form meanings that the juxtaposition of happenings can inform, if using one's intuitive powers.

Insight

The cliché of the "Aha!" the insight, the epiphany, the sudden answer to the problem brewing, the spectre of Archimedes rising from his bathtub and running down the street shouting "Eureka! I've got it!" about the theory of the displacement of water, all engage. Writers also experience insight, which is a well-researched construct with psychological experiments and studies (cf. Sternberg & Davidson, 1995). Even the ironic and intellectual David Shields in his ground-breaking manifesto (2010) confessed to having an insight:

> I'm not a big believer in major epiphanies, especially those that occur in the shower, but I had one nearly twenty years ago and it occurred in the shower: I had the sudden intuition that I should take various fragments of things—aborted stories, outtakes form novels, journal entries, lit crit— and build a story out of them. ... The initial hurdle (and much the most important one) was being willing to follow this inchoate intuition, yield to the prompting, not fight it off, not retreat to SOP. (pp. 172–173)

As this quote and the insight studies show, insight, while it has the appearance of suddenness, really requires the preparation of hard work, and is a reconceptualization of an old problem with new information added, and helps with solving ill-structured problems.

Nobel prize-winner Gabriel Garcia Marquez, a newspaper reporter at the time, drove between Acapulco and Mexico City for his work. One day, he had an insight about the first chapter of his 1967 novel, *One Hundred Years of Solitude*. When he got home, he cautioned his family he did not want to be disturbed, and he went to his room and stayed there for about 10 hours a day for 18 months and wrote the magical realist novel (Retrieved from newsletter@americanpublicmedia.org, March 7, 2009. *The Writer's Almanac*).

Imagery

Imagery precedes imagination; it is the sensory pictures (visual, auditory, gustatory, tactile, smell) by which imagination tells stories. Poet W. S. Merwin said that "Real image is a kind of fusion of all aspects of perception and of being in one tiny focus … you can't make images, you can't sort of calculate an image … you can steal almost anything, but you can't steal images" (*Contemporary Authors New Revised*, 15, p. 325). Amy Tan (1993) fused dreams, imagination, and image in discussing her creative process. "The kind of writing I do is very dreamlike. The process I go through is similar to what happens when I dream … when I'm writing I focus on a specific image and that image takes me into a scene" (p. 284). She concluded that "the kind of imagination I use in writing, when I try to lose control of consciousness, works very much like dreams. The subconscious takes over and it's fun" (p. 284).

Imagination

Imagination is a mental faculty whereby one can create concepts or representations of objects not immediately present or seen. There are two types of imagination: reproductive imagination which recreates memory and productive imagination which forms concepts that are derived from objects. The *Oxford English Dictionary* (OED) defines imagination as "the creative faculty of the mind in its highest aspect; the power of framing new and striking intellectual conceptions; poetic genius."

Using imagination is stimulating and entertaining. While your author wrote my first novel, *The Three-Week Trance Diet* (1985) I chuckled as the absurd happenings moved out of my imagination to my fingers onto the IBM Selectric keyboard. I was seeing things—a fat girl named Bwana crouching in the corner of her mother's condo taking one Brussels sprout and chewing each leaf thirteen times while her mother and her mother's young lover were in the next bedroom. Each day I wrote ten pages double-spaced. After writing throughout the morning and afternoon, I took a walk through our town in the evening, and I imagined what my characters would do during my writing session tomorrow. I was improvising and I felt like I was creating jazz. One of the reviewers said it had "zest, fun." Another called it a "postmodern satire." When it won a first novel contest over 80 other first novels, I was shocked and pleased. Such glee in imagination was described by writer Larry McMurtry: "I am more and more convinced that the essential reward of writing fiction is in the delight of seeing what you can make out of the sole tools of your imagination and your experience" (*Contemporary Authors New Revision*, p. 255).

Robert Olen Butler, the National Book Award-winning novelist and short story writer said that associations feed his imaginative work. One of his short stories came from a thread about Vietnamese boys and their trained crickets in one of his six unpublished novels. "Within 24 hours of writing that story, I had two dozen more story ideas. All these Vietnamese characters' voices began to present themselves to me. In the next year … I wrote one story after another" (Kelsey, 1996, p. 45). Robert Olen Butler is still writing novels, and the Vietnam vein is still yielding. On August 22, 2015, he posted on Facebook,

As the light grows dim outside, and to the sound of distant thunder, yes I am done. *Perfume River* done. Reading over is still to come, over and then over once more, tweaking and fiddling and refining and buffing, but done. Done. My best novel, my best book, I think. All that I am after 70 years, certainly. But in a few weeks all that I will be will be more than I am now, and a new book will begin.

This illustrates that often the very act of writing, the physical act of it, is important and vital to fueling the imagination in the creative process in writers.

In the words of Michael Chabon's novel-writer narrator Grady Tripp in the novel *Wonder Boys* (1995), writers often have too much imagination and must try to limit the productivity minds. Grady has a novel that is much too long:

> The problem, if anything, was precisely the opposite. I had too much to write: too many fine and miserable buildings to construct and streets to name and clock towers to set chiming, too many characters to raise up from the dirt like flowers whose petals I peeled down to the intricate frail organs within, too many terrible genetic and fiduciary secrets to dig up and bury and dig up again, too many divorces to grant, heirs to disinherit, trysts to arrange, letters to misdirect into evil hands, innocent children to slay with rheumatic fever, women to leave unfulfilled and hopeless, men to drive to adultery and theft, fires to ignite at the hearts of ancient houses. (Kindle location, pp. 143–151)

Incubation

In the creative process, incubation takes place while the creator is at rest, and is not thinking consciously about the problem. The unconscious is at work, and beneath the surface, a solution is bubbling up. Wallas (1926) called incubation one of the stages in his creative process of problem solving. Segal (2004) showed that people solve a problem more successfully if they go away and think about it rather than trying to solve it when it is presented.

Creators often incubate while driving, sleeping, exercising, even showering, as in the example from Shields, above. Incubation is a process that requires persistence, patience, and a trust that the preparation work has planted a seed and with waiting, contemplation, rest, and a relaxation of focus, the work will continue at a higher level. The answer will come. Naps are especially favored by some creators as a means of incubation. This part of the creative process might have led to the "nap rooms" in some Silicon Valley technology incubators. At the *Huffington Post*, writers can reserve the nap room. The online newspaper considers workday naps an important productivity enhancement. Writer Amanda Chan said, "I've used the nap room here for resting my eyes for a few minutes and for taking full-on power naps" (para. 6). Retrieved from https://sleep.org/articles/sleeping-work-companies-nap-rooms-snooze-friendly-policies/.

Improvisation

Improvisation seems to be a key part of the creative process and that is true in writers as well. Poet Hayden Carruth (1983) stated that writing is like playing jazz. "My best poems have all been written in states of transcendent concentration and with great speed... I have interfused thematic improvisation and ... metrical predictability" (p. 36). His poems use meter and rhyme that he disguises within the line so that the reader will not notice. "Jazz gives us a new angle of vision, a new emphasis ... in creative intuition" (p. 35).

The poet James Merrill (1992) used automatic writing as an improvisational technique: "Writing down whatever came into one's head, giving oneself over to every impulse—reasonable and unreasonable—concrete and abstract...is a means of granting oneself permission to speak from the heart, the depths of one's unconscious, the edges of the language" (p. 97). Other writers worked similarly. William Butler Yeats used both his own and his wife's automatic writing as inspiration for work. Poet Octavio Paz also engaged in the practice.

General Practices in the Creative Process

Need for Solitude

Solitude is necessary in order to focus on the inner self and its creative expressions. The granddaddy of all famous writers on solitude, Henry David Thoreau, said this, which is, of course, no surprise: "I find it wholesome to be alone the greater part of the time. To be in company, even with the best, is soon wearisome and dissipating. I love to be alone. I never found the companion that was so companionable as solitude" (Thoreau, 1845, Ch. 5). Poet and novelist Jim Harrison (1991) wrote that he requires a few months of near solitude every year: "I have learned … that I must spend several months a year, mostly alone, in the woods and the desert in order to cope with contemporary life, to function in the place in culture I have chosen." He walks and loses his "lesser self" in the "intricacies of the natural world." In nature, he is able to imagine himself back to 1945, and "the coyotes, loons, bear, deer, bobcats, crows, ravens, heron and other birds that helped heal me then, are still with me now" (p. 317).

Meditation

Meditation is increasingly within the public consciousness. Some have said the past decade and the next are marked by a spiritual need—a need for communion with the self and with others. An ongoing curiosity about eastern religions continues from the 1960s. Writers, artists, and scientists with backgrounds in Christianity and Judaism seem to reject dogma and they also reject mystical traditions they find in their home grown religions. They begin to study eastern religions.

A surprising number of writers, for example, have embraced Buddhism. One suspects this is because of the attention paid to meditation, to solitude, to the going within oneself of that religious faith. Here is a partial list: Allen Ginsberg, Robert Bly, W. S. Merwin, Anselm Hollo, Anne Waldman, Gary Snyder, Jane Augustine, John Cage, William Heyen, Lucien Stryk, and Philip Whalen (Johnson and Paulenich, 1992). Poet and songwriter Leonard Cohen spent several years in a Buddhist monastery (Simmons, 2012).

Others have embraced the contemplative life of the Christian monastery—for example, poet and minister Kathleen Norris and poet and priest Daniel Berrigan. Judaism is also explored through meditation. Dani Shapiro published a book on her spiritual journey through meditation, yoga, and back into her orthodox Jewish roots and the study of Kabbalah (Shapiro, 2010). The vehicles for discovering one's self are breath control, meditational technique, yoga or other eastern practice, visualization, imagery. Often the creative work follows the meditation, and the meditation is a preparatory ritual for the creative work. Meditation feeds the spiritual nature of writing.

Exercise

Walking seems to be the exercise of choice among writers. In Hodges (1992), the walking habits of several British writers were discussed. Samuel Johnson liked to make up his words while walking in the park. Coleridge thought about what he was going to write as he walked on bumpy terrain. Wordsworth paced back and forth on flat gravel. Wordsworth's sister Dorothy would walk with him on rambles through the British and Scottish countryside, and she would take dictation from him. The film, *Pandaemonium* (O'Hagan & Temple, 2001) shows imagined images of Wordsworth and Dorothy walking after they visited Coleridge. Tennyson would try out his latest poems while walking. Dickens walked around and around his house during the day, hitting himself on the forehead. He also was likely to take walks during the night. A. E. Chesterton, J. M. Barrie, Stephen Potter, and the Brontë sisters also circled the table where they wrote as they thought about what they were going to write in their novels. Hodges (1992) who collected these anecdotes, called it "ritualistic pacing" (p. 216).

American writers also walk as a ritual while writing. Upton Sinclair (1962) said he wore a path six inches deep in the woods near where he composed one of his novels one summer. Poet Robert Watson (1992) described how he wrote while walking:

> In those days and for years and years later I composed mostly on foot, where the rhythm of walking fed the rhythm of my poems. Usually I carried a pencil and used envelopes in my pockets so that I would write down the lines that came to me, often on the hood of a parked car. No poem ever flew into my mind that was complete and didn't need extensive repairs. (p. 160)

Poet and writer Michael Bugeja (1992) compared himself to Robert Frost: "I could not compose poems in our house, scene of bickering and discontent, so would walk like Robert Frost, an early influence, pacing out meter as one paces out a life" (p. 301). As with the other themes, there are many other examples. For example, see Piirto (2011a).

Norwegian writer Karl Ole Ausgart in an interview with J. Barron (2013) thought that muscle memory was able to inspire. When one does a certain exercise that is in the body, one recalls all those experiences:

> Writing is recalling. In this matter I am a classic Proustian. You're playing football for the first time in twenty years, for example, doing all those movements again, and it makes the body remember not only the strangely familiar movements, but also everything connected to playing football, and for some seconds, a whole world is brought back to you. Where did it come from? I think that all our ages, all our experiences are kept in us, all we need is a reminder of something, and then something else is released. (Barron, 2013, para. 16)

The writing process in creative writers thus involves the physical.

The Creative Process as Study

A major part of any creative process is study of the domain. Many people want to learn to write fiction, poetry, plays, and the like. There are many programs at universities, taught by graduates of similar programs. Many students major in creative writing both as undergraduates and graduates. Degrees are usually Bachelor of Fine Arts, Master of Fine Arts, and Ph.D. in Creative Writing. The curricula for these engage the writers in exercises and in drills as well as in workshops where class members' works are dissected by other class members. Most advertisements in *The Chronicle of Higher Education* for teachers of creative writing have, as their requirements, the possession of an MFA or Ph.D. in Creative Writing. The days where a teacher of creative writing could possess a B.A. or a M.A. in English or a Ph.D. in English and be a creative writing teacher are past. The value of studying how to write creatively was asserted by Robert Love Taylor (1993):

> To my delight, I found that you could take a course called creative writing. Handing in for assignments the poems that I would be writing anyway, I thought I was getting away with something. But in Daniel Knapp, who taught the introductory creative writing courses, I had a reader who combined sympathy and judgment. If my friends mainly applauded me, Daniel Knapp—and later, Kay Boyle—more soberly showed me how much I had yet to learn. (p. 263)

Ritual

A ritualistic approach to the act of writing is common among writers. Ritualistic practices vary. Currey (2013) published a whole book called *Daily Rituals: How Artists Work*. Some writers ritualistically read before writing. Poet and novelist Cynthia Ozick said, "I start priming the pump which is to read ... I read in order to write" (Teicholz, 1988, pp. 198–199). Poet Jeffrey Harrison described it thus: "I was reading a book of poems ... and it was reading those poems and being in a resulting poetic frame of mind ..." (Biles, 2015, para. 10).

Anne Rivers Siddons described her ritual when she begins a novel: "I have very tough, strict little rituals, especially when I'm just getting started to write a book. Once I'm under way the ritual is just repetition. I need to do it exactly the way I did it before to keep writing" (Siddons, 1993, p. 141). She makes nests from papers and places them throughout the house. She is distracted and has been known to walk into walls. "I once put my kitten in the refrigerator and put the orange juice carton out the back door." She fixes a big breakfast, cleans the house and goes to her writing studio." She is gone for three or four days until the book is on its way. James Hall's (Hall, 1993) ritual is "sort of warming up my fingers on the keys, just writing automatically." He pulls out a previous manuscript and starts monkeying with it. "The actual movement of my fingers across the board and hearing the characters start to talk is what" has him begin to write (p. 112).

The Influence of the Transpersonal: The Mystical Mystery

The creative process as described by many writers is not a cognitive process with a set sequence. In fact, the writers are often struck by the mysteriousness of how the work emerges, and they struggle to find logical explanations, but instead, they have to resort to eloquent descriptions with symbolism and metaphor.

- E. L. Doctorow: "It's like driving a car at night. You never see farther than your headlights, but you can make the whole trip that way" (Plimpton, 1988, p. 305).
- Scott Russell Sanders: "When I was in the flow of work, I felt free and whole. I played language as a pianist plays music, my fingers and ears captured in it, my body swaying" (Sanders, 1992, p. 248).
- Dick Allen: A sense of mysticism, a complete dissolving into wonder and beauty has been with me through my life. I remember always feeling nearly ecstatic in childhood (*Contemporary Authors Autobiography Series, 11*, p. 4).
- Octavio Paz: I wrote the first thirty verses as if someone were silently dictating them to me. I was surprised at the fluidity with which those hendecasyllabic lines appeared one after another. They came from far off and from nearby, from within my own chest. Suddenly the current stopped flowing. I read what I'd written. I didn't have to change a thing (MacAdam, 1991, p. 113).
- William Styron: When I'm writing at my best, I'm aware that I'm tapping subconscious sources (Styron, 1993, p. 276).

Wonder. Beauty. Dissolving. Disintegrating. Fluidity. What language to be used by those who treasure precision in language. This mystical sense that is described here has, for centuries, been described as having something in common with the divine.

Novelist Mark Helprin described an experience when he was young and injured from a fall and wandering through Europe, ending up in Paris, finding a hotel room which smelled of burning rubber, sleeping, and waking up free of pain. "In the absence of pain, I began to write on the blotter. As I did so, I lost track of what was up and what was down. I seemed to be revolving in space, without either gravity or time" (Linville, 1993, p. 186). In the morning he found he had written several paragraphs on the Hagia Sophia where he had never visited. "I felt such satisfaction and joy that I knew I had found my profession for life" (p. 186). The quality of the writing had "force, compression and beauty" and he was "stunned." Helprin stated, "Since then, the feeling has been much the same whenever I put a pen to paper. I lose the world, I concentrate so deeply that hours pass in an instant (for years I used to think that something was wrong with my clocks), and I emerge calm and strong" (p. 186). Csikzentmihalyi (1990) used the term "flow" (which was also used by Barron (1968)—see quote above) to describe this experience; Ghiselin called it "oceanic consciousness" (1952).

Discussion

A whole book could be written by this writer on the creative process in writers, as an example after example for theme after theme tumbles from her reading and research. Perhaps I will do so. In the interim, this essay has demonstrated the multifaceted nature of the creative process in writers; it is varied, bright, and yields much fodder for thought to those who study the real words and accounts of the writers themselves.

Caveat: The glossing of the material into the themes and subthemes is arbitrary and themes could be added or subtracted, condensed or expanded. When a qualitative researcher considers how to present her research, the decisions about themes are important but may be flexible and may change. For example, I added Tolerance for Ambiguity to the Five Core Attitudes later (in 2006) than in my 2004 work. One Core Attitude was called Naiveté, but it is now solidly called Openness to Experience, with the more recent studies utilizing the NEO-PI-R.

References

Barker, C. (1993). Clive Barker. In N. Epel (Ed.), *Writers dreaming* (pp. 31–42). New York, NY: Carol Southern Books.

Barron, F. (1968). *Creativity and personal freedom.* New York, NY: Van Nostrand.

Barron, F. (1995). *No rootless flower: An ecology of creativity.* Cresskill, NJ: Hampton Press.

Barron, J. (2013, December 26). Completely without dignity: An interview with Karl Ove Knausgsard. *Paris Review Daily.* Retrieved from http://www.theparisreview.org/blog/2013/12/26/completely-without-dignity-an-interview-with-karl-ove-knausgaard/.

Barthelme, D. (1981). *Sixty stories.* New York, NY: G. M. Putnam Sons.

Biles, J. (2015, July). Interview with Jeffrey Harrison: The Writers' Almanac. Retrieved from http://writersalmanac.org/bookshelf/jeffrey-harrison.

Brooks, J. (1995). Interview with Chinua Achebe: The art of fiction, CXXXVIV. *Paris Review, 133,* 142–166.

Bugeja, M. J. (1992). The muse at home. In E. Shelnutt (Ed.), *My poor elephant: 27 male writers at work* (pp. 295–310). Marietta, GA: Longstreet Press.

Carruth, H. (1983). The formal idea of jazz. In S. Berg (Ed.), *In praise of what persists* (pp. 33–44). New York, NY: Harper & Row.

Chabon, M. (1995). *Wonder boys.* New York, NY: Random House.

Contemporary authors autobiography series, 11, (CAA). (1990). Detroit, MI: Gale Research.

Contemporary authors new revised. (1981ff.). Detroit, MI: Gale Research.

Csikszentmihalyi, M. (1990). *Flow.* New York, NY: Cambridge University Press.

Currey, M. (2013). *Daily rituals: How artists work.* New York, NY: Alfred A. Knopf.

Daniels, J. (2012, June 6). Aaron Sorkin. *Interview.* Retrieved from www.interviewmagazine.com/film/aaron-sorkin/#page2.

Dee, J. (1988). Interview with John Hersey. In G. Plimpton (Ed.), *Writers at work* (pp. 99–136). New York, NY: Viking Penguin.

Faggen, R. (1995). Czeslaw Milosz: The art of poetry LXX. *Paris Review, 133,* 242–273.

Getzels, J., & Csikszentmihalyi, M. (1976). *The creative vision: A longitudinal study of problem finding in art.* New York, NY: Wiley.

Ghiselin, B. (Ed.). (1952). *The creative process.* New York, NY: Bantam.

Graham, J. (1995). *The dream of the unified field: Selected poems 1974–1994.* New York, NY: HarperCollins.

Guppy, S. (1989). Interview with Edna O'Brien. In G. Plimpton (Ed.), *Women writers at work: The Paris Review interviews* (pp. 337–359). New York, NY: Viking Penguin.

Hall, J. W. (1993). James W. Hall. In N. Epel (Ed.), *Writers dreaming* (pp. 106–118). New York, NY: Carol Southern Books.

Harrison, J. (1991). *Just before dark: Collected nonfiction.* New York, NY: Houghton Mifflin.

Hart, H. (2000). *James Dickey: The world as a lie.* New York, NY: Picador.

Hodges, J. (1992). *The genius of writers: A treasury of facts, anecdotes, and comparisons: The lives of English writers compared.* New York, NY: St. Martins Press.

John-Steiner, V. (1997). *Notebooks of the mind.* New York, NY: Oxford University Press.

Johnson, C. (1993). Charles Johnson. In N. Epel (Ed.), *Writers dreaming* (pp. 119–132). New York, NY: Carol Southern Books.

Johnson, K., & Paulenich, C. (Eds.). (1992). *Beneath a single moon: Buddhism in contemporary American poetry.* Boston, MA: Shambhala.

Joyce, J. (1922). *Ulysses.* Paris, FR: Sylvia Beach.

Kakutani, M. (1995). Interview with Woody Allen: The art of humor, I. *Paris Review, 133,* 200–222.

Kelsey, M. (1996, January/February). An interview with Robert Olen Butler. *Poets & Writers Magazine, 24*(1), 40–48.

Kevles, B. (1989). Interview with Anne Sexton. In G. Plimpton (Ed.), *Women writers at work* (pp. 263–289). New York, NY: Viking Penguin.

King, S. (1993). Stephen King. In N. Epel (Ed.), *Writers dreaming* (pp. 133–143). New York, NY: Carol Southern Books.

Langan-Fox, J., & Shirley, D. A. (2003). The nature and measurement of intuition: Cognition and behavioral interests, personality, and experiences. *Creativity Research Journal, 15*(2), 207–222. https://doi.org/10.080/104004 19.2003.9651413.

Linville, J. C. (1993). Interview with Mark Helprin: The art of fiction, CXXXII. *Paris Review, 126,* 160–199.

Lubart, T. (2009). In search of the writer's creative process.In S. B. Kaufman & J. C. Kaufman (Eds). *The psychology of creative writing* (pp. 149–165). New York, NY: Cambridge University Press.

MacAdam, A. (1991). Interview with Octavio Paz. *Paris Review, 119,* 82–113.

MacKinnon, D. (1978). *In search of human effectiveness: Identifying and developing creativity.* Buffalo, NY: Creative Education Foundation.

Marquez, G. G. (1967). *One hundred years of solitude.* New York, NY: HarperCollins.

Maslow, A. (1998). *Maslow on management*. New York, NY: John.

May, R. (1975). *The courage to create*. New York, NY: Bantam.

McCullough, J., & Simpson, M. (1994). Alice Munro, the art of fiction no. 137. *Paris Review, 137*. Retrieved from http://www.theparisreview.org/interviews/1791/the-art-of-fiction-no-137-alice-munro.

Merrill, J. (1992). Permission to speak. In E. Shelnutt (Ed.), *My poor elephant: 27 male writers at work* (pp. 83–100). Marietta, GA: Longstreet Press.

Moyers, B. (1995). *The language of life: A festival of poets*. New York, NY: Doubleday.

Mukherjee, B. (1993). Bharati Mukherjee. In N. Epel (Ed.), *Writers dreaming* (pp. 160–166). New York, NY: Carol Southern Books.

Nichols, J. (1993). John Nichols. In N. Epel (Ed.), *Writers dreaming* (pp. 178–187). New York, NY: Carol Southern Books.

O'Hagen, N. [Producer] & Temple, J. [Director]. (2001). *Pandaemonium* [Motion picture]. London, UK: British Broadcasting Corporation.

Phillips, R. (1989). Interview with Joyce Carol Oates. In G. Plimpton (Ed.), *Women writers at work: The Paris Review interviews* (pp. 361–384). New York, NY: Viking Penguin.

Piirto, J. (1985). *The three-week trance diet*. Columbus, OH: Carpenter Press.

Piirto, J. (1998). *Understanding those who create* (2nd ed.). Tempe, AZ: Great Potential Press.

Piirto, J. (2002). *"My teeming brain": Understanding creative writers*. Cresskill, NJ: Hampton Press.

Piirto, J. (2004). *Understanding creativity*. Scottsdale, AZ: Great Potential Press.

Piirto, J. (2007). Creativity. In J. L. Kincheloe & R. A. Horn (Eds.), *The Praeger handbook of education and psychology* (pp. 310–320). Santa Barbara, CA: Greenwood Press.

Piirto, J. (2008). Rethinking the creativity curriculum: An organic approach to creativity enhancement. *Mensa Research Journal, 39*(1), 85–94.

Piirto, J. (2009). The creative process as creators practice it: A view of creativity with emphasis on what creators really do. In B. Cramond (Ed.), *Perspectives in gifted education: Creativity* (pp. 42–67). Denver, CO: Institute for the Development of Gifted Education, University of Denver.

Piirto, J. (2010). The five core attitudes and seven I's for enhancing creativity in the classroom. In J. Kaufman and R. Beghetto (Eds.), *Nurturing creativity in the classroom* (pp. 142–171). New York, NY: Cambridge University Press.

Piirto, J. (2011a). *Creativity for 21st century skills: How to embed creativity into the curriculum*. Rotterdam, The Netherlands: Sense Publishers.

Piirto, J. (2011b). Synchronicity. In M. Runco & S. Pritzker (Eds.), *Encyclopedia of creativity* (2nd ed., Vol. 2, pp. 409–413). San Diego, CA: Academic Press.

Plimpton, G. (1988). Interview with E. L. Doctorow. In G. Plimpton (Ed.), *Writers at work* (pp. 299–321). New York, NY: Viking Penguin.

Plimpton, G., & Crowther, E. (1988). Interview with E. B. White. In G. Plimpton (Ed.), *Writers at work, Eighth Series* (pp. 2–23). New York, NY: Penguin Books.

Policastro, E. (1999). Intuition. In M. Runco & S. Pritzker (Eds.), *Encyclopedia of creativity* (Vol. 2, pp. 89–93). San Diego, CA: Academic Press.

Price, R. (1993). Reynolds Price. In N. Epel (Ed.), *Writers dreaming* (pp. 200–208). New York, NY: Carol Southern Books.

Sanders, S. R. (1992). Letter to a reader. In E. Shelnutt (Ed.), *My poor elephant: 27 male writers at work* (pp. 236–254). Marietta, GA: Longstreet Press.

Segal, E. (2004). Incubation in insight problem solving. *Creativity Research Journal, 16*(1), 141–148.

Shapiro, D. (2010). *Devotion: A memoir.* New York, NY: HarperCollins.

Shields, D. (2010). *Reality hunger: A manifesto.* New York, NY: Random House.

Siddons, A. R. (1993). In N. Epel (Ed.), *Writers dreaming* (pp. 237–243). New York, NY: Carol Southern Books.

Simmons, S. (2012). *I'm your man: The life of Leonard Cohen.* New York, NY: HarperCollins.

Sinclair, U. (1962). *The autobiography of Upton Sinclair.* New York, NY: Harcourt, Brace, and World.

Solzhenitsyn, A. (1963). *One day in the life of Ivan Denisovich* (R. Parker, Trans.). New York, NY: Dutton.

Sternberg, R., & Davidson, J. (Eds.). (1995). *The nature of insight.* Cambridge, MA: MIT Press.

Styron, W. (1993). In N. Epel (Ed.). *Writers dreaming* (pp. 270–280). New York, NY: Carol Southern Books.

Tan, A. (1993). In N. Epel (Ed.), *Writers dreaming* (pp. 281–288). New York, NY: Carol Southern Books.

Taylor, R. L. (1993). The territory of memory. In E. Shelnutt (Ed.), *My poor elephant: 27 male writers at work* (pp. 255–268). Marietta, GA: Longstreet Press.

Teicholz, T. (1988). Interview with Cynthia Ozick, 198. Reprinted in G. Plimpton (Ed.), *Writers at work, eighth series* (pp. 195–223). New York, NY: Viking Penguin Books.

Thoreau, H. D. (1845). *Walden.* Retrieved from http://thoreau.eserver.org/walden00.html.

Wallace, D. F. (1996). *Infinite Jest.* New York, NY: Little, Brown.

Wallace, I. (1962). *The prize.* New York, NY: Dutton.

Wallace, I. (1968). *The writing of one novel.* New York, NY: Simon and Schuster.

Wallas, G. (1926). *The art of thought.* New York, NY: Harcourt Brace Jovanovich.

Warner, M. (1989). Interview with Rebecca West. Reprinted in G. Plimpton (Ed.), *Women writers at work: The Paris Review interviews* (pp. 71–105). New York, NY: Viking Penguin.

Watson, R. (1992). Luck and chance. In E. Shelnutt (Ed.), *My poor elephant: 27 male writers at work* (pp. 153–165). Marietta, GA: Longstreet Press.

Wilmer, C. (1995). Thom Gunn: The art of poetry, no. 71. *Paris Review,* 135. Retrieved September 17, 2015, from http://www.theparisreview.org/interviews/1626/the-art-of-poetry-no-72-thom-gunn.

Wood, W. C. (1988). Interview with Robert Stone. Reprinted in G. Plimpton (Ed.), *Writers at work* (pp. 343–375). New York, NY: Viking Penguin.

5

Collaborative Scriptwriting: Social and Psychological Factors

Samira Bourgeois-Bougrine and Vlad Petre Glăveanu

The script as a collaborative creative product is of utmost importance for the film industry; it is the blueprint, the heart and the backbone of the movie. A film project may originate from a screenwriter, as described in the "script development" stories of major Hollywood studios, or it can start with a demand, i.e. a director or producer can commission a scriptwriter, novelist or another director to write or co-write the script (Fergusson, 2004). In either of these cases, writing appears as a "solo work, authored by a lone genius; it unfolds in collaborative circle" (Sawyer, 2009, p. 174). The argument put forward in this chapter is that, whether based on a concept, an idea, a true story, an existing script, a book or a synopsis, writing a script for a movie is in fact a deeply collaborative creative process. It involves iterative phases of private writing

S. Bourgeois-Bougrine (✉)
Paris Descartes University, Paris, France
e-mail: samira.bourgeois-bougrine@parisdescartes.fr;

V. P. Glăveanu
Department of Psychology, Webster University, Geneva, Switzerland

© The Author(s) 2018   **123**
T. Lubart (ed.), *The Creative Process*, Palgrave Studies in Creativity and Culture,
https://doi.org/10.1057/978-1-137-50563-7_5

of drafts and collaborative interactions to analyse these texts and provide feedback that leads to a new draft. The process of rewriting several versions of a script seems to be universal among screenplay writers, engaging often the director and the producer of the film as well as actors and other members of the film crew. The simultaneously individual and social working contexts foster interactions with others that can be, at times, a source of tension and frustration (Condor, 2010). Ultimately, a final version of a script is virtually impossible as scripts are often modified before and during the shooting, as well as during the movie editing stage.

Although a considerable amount of research in psychology is dedicated to creative writers, screenwriting itself has received little attention (Pritzer & McGarva, 2009). This may be due to the fact that the screenplay writer's activity is rarely visible, its' end product, the script, is almost never seen by audiences, a symptom of the general marginalisation of screenplay writers within a highly hierarchical film industry (Condor, 2010; Pritzer & McGarva, 2009). Psychologists are generally interested in understanding the mental processes involved in the solo writing phases. However, a complete explanation of the processes that result in a final text requires a thorough analysis of the influence of the social interactions before and between private writing phases (Swayer, 2009, p. 174). Yet, the simultaneous study of the social interactions and mental processes in screenplay writing is rare.

The present chapter aims, in this context, to contribute to our understanding of the social and psychological factors involved in creative writing in the case of both recognised French screenwriters and students of this profession. It is hoped therefore to add not only to the scarce literature on screenplay writing in psychology but, more broadly, to our understanding of particular social and psychological factors and processes involved in this activity. To achieve this, our study focused first on the analysis of interviews of twenty-two French award-winning professional scriptwriters who presented the general way in which they create a new script and illustrated these aspects by describing one of their most successful productions. The second part aimed at understanding how a group of six students of scriptwriting and film-making studies learn the process of writing a script. They had eight weeks to create a script starting from a common theme: "A 19 year old women was found dead, murdered by eight knife stabs, in the nave of Notre Dame". Students completed a booklet, which consisted of a self-report part, focusing on the stages of the creative process, as well as an open part in which they were able to record their weekly progress.

Social Factors and Processes in Screenplay Writing

As any collaborative task, writing a script requires establishing and maintaining a shared vision, a common understanding and distributed knowledge within the creative team. Social interactions with the sponsor (in these cases a film director or producer) occur before the actual scriptwriting commences, meant to evaluate a demand or proposal as well as to develop relations based on empathy, to get to know each other and assess the feasibility of the endeavour. During the cycle of writing and rewriting, discussions with friends, producer or director foster inspiration, help solve problems and overcome creative blocks.

Illustrating the collaborative nature of creative endeavour, one of our participants noted that: "Speaking more generally, an artwork is largely a 'delivery'. In the film industry, it is a 'delivery' involving lots of midwives as well as surgeons and hospital staff. Producer, technicians, writer, etc., should help a director to give birth to an idea dear to his heart" (S18).

Diversity and Social Cohesion

Throughout their careers, our participants moved from one film project to another and were involved in a total of 373 films within a wide spectrum of productions: author, drama, adaptations for TV, comedy or documentary. They have different backgrounds: a degree in scriptwriting, film-making, literature, psychology, history, journalism, philosophy or political sciences. Some writers were both scriptwriters and novelists, others directors writing scripts for their own films and co-writing other directors' films.

For the majority of the professional writers interviewed, writing a script had its origin in a demand or proposal mainly from a director and sometimes from the producer of a TV series or documentary to develop an original idea, adapt a book for the screen, or modify an existing script. Some of them worked regularly with the same directors, actors or producers and came to develop considerable tacit knowledge regarding their work habits, which facilitates communication and social cohesion. Three participants who regularly work with a director commented on this tacit knowledge:

> (name of the director), with whom I worked the most, it's someone who sees the script as a kind of intellectual question rather than a story to tell. So there are days and days of discussion together, debate, theorizing, shared readings, etc. (S15)
>
> For example [name of the director] likes "putting" a first layer of dialogues at that time [structuring phase-outline], from which generally not much remains in the end, but he needs to go through it. (S22)
>
> For example, I wrote for [name of a director] and I've come to understand that she does not film spaces at all... she prefers people... When directing she always forgot to film spaces, scenery, and location, that have an important role in the script. Everyone has his style... [name of another director], I know she does not like to film only objects, for example. She does not like when there is no action. (S5)

It is widely believed that "tacit knowledge of one's teammates' work habits matters for successful performance" and the accumulation of this knowledge is possible only through the experience of working with each other (DeVaan, Vedres, & Stark, 2014).

In other cases, where there is no previous experience of working together, writers need to spend time to get to know each other, to discover the habits and the filming preferences of the director, as mentioned in the following interview passages:

> For me, it is essential to discuss the options for the film. If you write things that the other doesn't want to film or doesn't know how to film, even if the script is good, it will be a catastrophe on the screen... (N5)

And this is also part of the magic of the meeting with the director, we discover every time what his own tastes are, what his "universe" is, to which 'family' in the cinema he belongs to… For me, it is important to both respect this 'family' and at the same time try to introduce him to something else, to make him discover places where he would not have gone alone… (S2)

When I meet someone who has expressed his desire to work with me, I take the time to know if I would want to work with him… This means that we start with preliminary meetings, where we make casual talk about this and that, about life, not necessarily the subject, but things that help to know each other. If he is someone very hysterical who tends to worship an idea the night before only to burn it the next day, there is no way it can work. If it's someone who knows too much what he wants, it also bothers me because I feel I will just have to fill the gaps. (S10)

Participants commented on bad experiences with some directors they qualified as "psychopaths" or "lunatics", people who engage in a sort of "sado-masochistic" or "vampiric" relationship with them. The writers, on the other hand, seek to have a cordial relationship based on trust and esteem. According to Fergusson (2014), professional creators, such as screenplay writers, develop different practices to achieve "security, sanity, harmony, respect and success in their creative collaboration with motion picture development team members" (p.).

Dialogue and Openness

After accepting a demand for writing a script, the writer tries to under-stand the "heart" of the project through discussions with the sponsor of the script, the director, producer or co-writer. These meetings are frequent (usually on a daily basis) and are associated in general with the feeling of interest and pleasure in sharing ideas. Dialogues are free and open and these discussions are of great importance for the majority of the scriptwriters. In this phase, there is a lot of "why…", "I would like…", "what if…" and openness to whatever comes. The openness, the trust and the pleasure in sharing ideas and desires that characterise these meetings might lead to a positive mood and a feeling of freedom to explore unconventional ideas about the main characters and the story to be told. Research suggested that the feeling of being free from constraints (Steidle & Werth, 2013) and in a positive mood (Labro & Patrick, 2009) promotes creative idea generation. Events happening in the distant future, for example, the project of the film in our study, are represented in a more abstract, structured, high-level manner. According to Construal Level Theory, a framework that links processing styles and psychological distance (Liberman, Trope, McCrea, & Sherman, 2007), processing information in a global, abstract and explorative way helps in finding creative solutions and ideas (Förster & Dannenberg, 2010; Steidle & Werth, 2013).

One of the major aims of these meetings is to understand the motivation behind the film and the personality of the characters. Some writers described these meeting as "psychoanalysis" sessions, necessary as well to discover the motivation and preferences of the director:

I think that working with the director would count as psychoanalysis … we talk about 'sessions'. We meet every day – this is a serious analysis… Especially when it's personal… we have to talk about very intimate things, otherwise it will not work. Sometimes you're dealing with people who have difficulty verbalizing, and you're there to listen, bring them slowly to speak about themselves. (S13)

The first step is really the dialogue. There is lot of talk and dream. Often the question of the characters comes first. As co-writer, I try to understand what the director is looking for. So I get him to talk, I listen to him, I try to figure out why he has selected a particular theme, what is the meaning of the character and in which universe he wants him to evolve. (S20)

We always start by talking about the characters. This is my method. Speaking about characters, means imagining, for each, their history, their background as they say, their past: where they come from, what is their social background, what happened to them, etc. We imagine. Even if it means, not using it at all… We even had a kind of identification cards with the stories of each character… And I am convinced that the accuracy of that (stories about the characters), although paradoxically it must disappear in the film, helps give the characters the complexity, thickness, opacity, contradictions… because a character that would be the same from the beginning to end offers no interest for me. (S10)

I have to strike a deep chord… he (the director) must explain to me in an analytical way or another, whatever. But, I need to know why he is interested in this subject. Because if I do not know, I might take him to side roads, and this is not the desired outcome… (S18)

A scriptwriter offered an interpretation of the cause of a bad script as a lack of understanding the meaning of the story rather than the writing:

Because a very bad scenario, it is not that it is not well written, but the question as to why this story has to be told has not been asked… when the purpose is lacking, in fact… When there is a purpose, a real drive, something will always come out of it… Because once we found it, we can write the outline in a month, and in three months we get a version one. It is better to go through it from the start, rather than make a first version within four or five months to put it in the bin and start asking the questions that should have been asked before. (S22)

Inspiration and Joint Problem-Solving

Depending on the project, the first version of the script could be written alone (when the director does not want to take part in writing) or with another writer or the director (writing with 'four hands' as mentioned by co-scriptwriters). In some cases, the writer had an advisory role, he/she reads what the director wrote and rewrote the text if needed. In other cases, the director asked the scriptwriter for a first version of the script and rewrote it afterwards:

He made me read it (a first draft of dialogues), and I usually take a lot of notes; it is covered with red, arrows, "yes," "no" and "that's great", "that's to be rewritten", etc. We meet, he validates my thoughts, and I leave with the text to rewrite it, to add another layer. I send it to him and then we meet, etc. At the end, it's a mille-feuilles [cream and vanilla layer French cake], we do not know who wrote what. (S22)

In the joint process of problem-solving, both directors and scriptwriters rely on each other's knowledge and opinions. Sometimes this blurs the line between their respective responsibilities. For example, scriptwriters need input from the director about the emotions the latter would like to convey or trigger:

> Above all, they (directors) generally have a clear idea of the emotion they want to create. Because in fact, what they tell you is the emotion they would like to have themselves as a movie viewer. So, inevitably it orients (the writing)… for example, I co-wrote with a director a script where there were trucks traveling at night. So I asked him: these trucks, how do you want to film them? What image do you see? Is it a helicopter view, or is the camera on a truck? Or is the camera on the side of the road and watching the trucks pass by? If one wants a grandiloquent style, the camera will be put on the helicopter, etc. According to what the director wants, grandiloquent or minimalist film, I have to adapt the writing. Each type of image conveys an emotion, which guides the writing… But directors do not like these kinds of questions. They feel that we step on their toes. (S11)

Writers often need someone to read or review their script and talk about it. This offers them a new inspiration and, most of all, creates the necessary distance between writer and work that allows the former to solve problems or observe new directions and possibilities for the script. Yet one needs to select a good adviser:

> There are good and bad readers. For example, [name of a director] is a very good reader and [name] who is a producer, can also be a good reader… what is very important to me is when a reader opens a new angle of the story we had not seen ourselves. Then it becomes interesting! Bad readers cling to details, say for example, 'but this girl, shouldn't she be blonde, rather than brunette?' With this, you cannot do much. (S12)
>
> Often, and this is the advantage of co-writing with a director, when one blocks, the other can find the solution. It can also be a third party, the producer for example… Once, with the director, we were both floundering; nose glued on our characters, we could not have the necessary distance… And it is the producer of the film, reading the script, working on it, who found the solution… (S20)

Collaboration in problem solving is ongoing. All the writers considered that the script is never finished and directors in general introduced changes before and even during the shooting. As the script progresses through the hands of many people, the scriptwriter could be asked to make changes just before or during the shooting. The suppression of some scenery or the rewriting of scenes could be related to a tight budget, shouting difficulties, or because the director, producer, sponsor (in the case of TV), actors, or technicians did not like a scene:

> Everyone has an opinion, and should be taken into account more or less. The director would say: 'My wife did not like the end', 'the sponsor does not want a child martyr'… And you are forced to constantly adapt. (S13)

> The producer will say for example: This character is not friendly enough. Give him a more sympathetic character. Maybe if he was a musician as well, or if he feeds the pigeons, etc. (S16)

> Yes, all the way through, there was debate about the end of the film: does the heroine leaves her husband or not? Personally, I thought she should not, and precisely, that was the story: a married woman meets a man, they have an affair, but she did not leave her husband. [name of a director], he has an opposite view… So in the end she left her husband. But up to the shooting and editing, there were really both sides. Everyone had his idea of an ideal ending. (S3)

> And after fifteen days of filming, [name of a director] said: "I am not able to shoot this scene". But, the scene in question was about 12 or 15 pages in the middle of the script. It's huge, and it was a pivot!… He knew he was in trouble and he wanted me to come. So I went. For a month, I commuted back and forth between home and the shooting locations… The story had been rewritten, but by different means. It was much narrower on two characters, but the others had to exist anyway, so we continued to ask questions, to move information, to rewrite. The actors, [three names], played the game, they were great. There was between all of us a great mutual trust. (S22)

Managing social interactions throughout the process of writing a script represents a key challenge as well as a key resource for the screenplay writer. In the process of co-writing and sharing knowledge, each party needs to understand the other's point of view while taking distance from one's own as a precondition for any successful collaboration: "*One should not be in the projection of the self in a scene so that, if one amends it, it's like a personal offence. Yet, many people are like that*" (S22).

Psychological Factors and Processes in Screenplay Writing

Up to this point, we have reviewed several social factors and processes that contribute to screenplay writing, in particular issues related to diversity and social cohesion, dialogue and openness, inspiration and joint problem-solving. In order to place them in the broader picture of scriptwriting activities, we will proceed by outlining the main stages of scriptwriting, as they appear from our interviews, and the psychological processes screenplay writers make reference to in each one of them. These phases are: impregnation, planning for action or structuring and production or the cycle of writing and rewriting the actual script. Two observations are required at this point. First, these phases are not seen as separate from each other but inter-related, with the possibility of returning to a previous phase, jumping ahead or engaging in more than one phase at a given time. Second, it is to be noted that we are not operating here with a strict separation between the psychological and the social. On the contrary, we conceive of these as interrelated and co-evolving facets of any human activity system (Cole, 1996) because social interactions and environments are both shaped by and actively shaping the psychological makeup and responses of the person, even in moments when the person is thinking or working alone (see Marková, 2003).

Impregnation

At the beginning of private writing episodes, participants commented on their difficulty to get started. Many of them mentioned how, at first, they collect a massive and usually disproportional amount of information, reading books, magazines, newspapers, consulting archives and photos, watching movies, etc. They will engage with these accumulated cultural resources in actual writing when a deadline is looming. Moreover, they need to expose themselves to a wide range of situations and cultural artefacts that have even a loose connection to the topic in a process they refer to as "turning around" their work:

> I like to hang around, read, do nothing… If I am the one in charge of writing the outline, I say: 'yeah, yeah, it's coming along' even though I have not written a f***g line. But I have taken some notes, I have read… I have worked internally. (S1)
>
> I am starting a new project and currently I am turning around. It's a kind of ritual: I am turning around the topic like a wolf around its prey. (S8)
>
> I find it hard to tackle the job head on. I first circle, I turn around. I have a relation to work a bit like the way I read the newspaper: when I spot a page in a newspaper that interests me, I first read all those with little interest to me to finish by reading the one that interests me. This is a very perverse mania; when I find a good idea, or a scene that I like, I turn around. Generally, I do a lot of things that are indirectly related to the work, I read a lot, I copy many texts that interest me, I see a lot of movies, I listen to a lot of music… I do many things around the work which I think will be the foundation of the work itself. (S4)

What this process of "turning around" a topic achieves is a form of "impregnation" with ideas related to the story being developed. As one of our participants aptly notes: "*one has to be really impregnated with his story… if one is sufficiently impregnated, ideas will emerge suddenly while on the bus… If you are not impregnated, nothing comes out, you have no idea…*" (S12)

In order to understand the psychological processes involved in this impregnation phase, we can refer to the Construal Level Theory framework. Liberman et al. (2007) showed that participants "would engage in an activity at a later point in time when it was described in abstract (rather than concrete) terms, when they had first considered why (rather than how)". Indeed, during the initial meetings with the sponsor, the writer considers the script project in terms of "why" rather than "how" and, just as in "talking therapy", s/he tries to understand the director's motivation behind the project of the film through daily discussions without censorship (see the section on Dialogue and openness). As suggested by McCrea, Liberman, Trope, and Sherman (2008), the mental association between level of abstractness and temporal distance is a bidirectional relationship: events that are distant in time tend to be represented more abstractly than events that are close in time and the level of representation of an event has effects on the time when the activity is performed. In other words, procrastination increases when thinking about the task in abstract terms rather than in concrete ones.

Writers put lots of emphasis on allowing the mind to wander freely, reflecting, thinking, daydreaming about the topic of the film and imaging situations, etc., before starting the actual writing. They made comments about how their brain works feverishly and continuously making unconscious associations and connections, how images emerge and how solutions are suddenly transferred to consciousness at a later stage of the creative process. They considered that incubation helps establish unconscious connections that are important for the productive phase of the creative process:

> Dreaming for as long as possible about things, in a chaotic and erratic way, I try to make this time of openness (last) as long as possible. I find that this space of connections made at the beginning, mainly because you are still not 'under the gun', is generally decisive. (S3)
>
> I need time to imagine, to leave some room for daydreaming… In any case, I think that it is like a washing machine cycle, you know, the ones that are very slow: we believe the machine is done but no, in fact it continues to work. In fact, you can do other things, even watch a silly TV show, there are neurons that connect and stay on the topic. It is foundational work being done: it moves, it is always there. (N13)

Interestingly, their description of what happens in their mind during this "passive" state is in line with the findings of neuroimaging studies of the brain (Andreasen, 2011; Andreasen, O'Leary, Cizadlo, Arndt, & Rezai, 1995): when individuals are given the instruction to "relax and simply think about whatever comes into their mind", the association cortices are the most active. These brain areas, which are referred to as the default network (DNT), are similar to those active during remembering in an autobiographical task. During this passive period, creative individuals (who have won prestigious awards in art and sciences) demonstrated stronger activations in DNT compared to control participants. As pointed out by Binder et al. (1999), such activations are adaptive: "by storing, retrieving, and manipulating internal information, we organize what could not be organized during stimulus presentation, solve problems that require computation over long periods of time, and create effective plans governing behaviour in the future". Subsequent research (Buckner, 2012; Buckner, Andrews-Hanna, & Schacter, 2008) showed that the activation of the DNT has been associated with: "constructing dynamic mental simulations based on personal past experiences such as used during remembering, thinking about the future, and generally when imagining alternative perspectives and scenarios to the present" (p.).

Planning for Action

This phase, which focuses on building the "architecture" of the story, includes the writing of a synopsis (the condensed version of the plot), an outline (a list of scenes and sequences) and a treatment (the elaborated version of the outline in 30–40 pages). The great majority of scriptwriters in our sample considered writing the synopsis before finishing the script to be an *"aberration due to production constraints"*; unless they are formally contracted to do so, they refuse to write it in the initial phase of the creative process because, as mentioned by one participant, *"It seems very artificial; it forces us to develop a story with characters we do not yet know… we have to pretend… I hate it"* (S2).

However, one director commented on the necessity for him to have a synopsis: *"When I write for myself and because I am not a famous director, I always have to 'go through' with a synopsis or treatment to get funding"* (S18).

The exact moment when the outline or treatment is written can vary. For example, an outline may be generated early in the process in order to understand the problem and then written formally before the production phase or after an initial version of the script. One major difference in this regard between scriptwriters was related to creating an outline and a treatment. Two groups can be identified based on this. A first group of ten scriptwriters considered the writing of an outline and/or a treatment as a daunting task and avoided it unless formally specified in the contract. They start the writing without a formal plan and seem to wander aimlessly or laboriously as reported below:

I'm going in all directions, like a hunter who wanders in a forest. (S16)

For an original script, I start from a situation, which seems interesting or funny. Then I try to see where it will lead me. In most cases, I don't know where I am going. Things build up gradually, and the meaning of all this emerges very late and sometimes not at all. (S6)

I start writing the scene, I know roughly where I want to go, even though I do not really know how to get there, and the fact of writing by following the character makes me find it. (S20)

For example, I start a scene, I try to write dialogues, and then I realize that it is too early. If you do not know where it all goes, you often find that the dialogues are very poor. So I always work everything at the same time: I write a synopsis and I start a scene… It is very mixed, and frankly it's a bit of a bloody mess. (S12)

I also write a lot of notes on small notebooks, about characters. Some things you will not see, that the public does not know. Although it will not directly serve the script, I can have fun to imagine… I enjoy imagining the characters 'off camera'… I do that at the beginning when I start and I am very anxious. There are cards everywhere: who was the grandfather of the heroin of the film? What was the love story of her mother? etc. All this eventually disappears, but I write about it a lot, even entire scenes so I do not know if they will end up or not in the script.

A second group of eleven scriptwriters (among them six with a specific education in scriptwriting) seemed very rigorous about developing an outline and treatment before starting to write: "*After this very open dialogue phase, I start with the treatment which I often offer to write alone. In general the directors agree, because the treatment is a step that pisses them off, while it allows me to already set things*" (S14). We will return to a discussion about these differences in our section on the education of screenplay writers. For the moment it is important to note that the process of writing a script is never a linear, "beginning to end" journey, but can take different paths and also involves a lot of "back and forth" movements (for a broader discussion of this activity as travelling a maze see Bourgeois-Bougrine et al., 2014). This is most obvious in the third phase of screenplay writing: producing the script.

Production

In contrast to the myth that creative ideas result from a "dreaming" person having a flash of insight, our participants insisted on the fact that insights, new ideas and solutions are the result of discipline and daily hard work, confirming also previous findings (Paton, 2012):

> It is necessary to be detached from any idea that could be close to 'inspiration', from all those fantasies you read about writers with their little notebooks where they would take a note of a sudden brilliant idea... If I were to write only when I am in the mood of writing, I would not go far... (S5)
>
> I'd say you must search, search, and again search; nothing comes without effort. Nothing is ever given to you. And this is also why regular daily work is required. There are days when nothing happens, and this is normal, so you should not blame yourself. And there are other days when things happen... and this is because for three days you have searched without finding anything... (S3)

They engage in work even when not inspired, focusing their attention on the task, developing ideas, solving problems and taking decisions or making choices between different alternatives. Participants described the writing process as a cycle, alternating between generating creative ideas and evaluating their potential through "a continual testing of and discarding of ideas as to their suitability for the story" (Nelmes, 2007, p. 111). Markman and Dyczewski (2010) described the portrait of an effective decision maker, problem-solver and goal pursuer as "an individual who displays the cognitive flexibility to think and act both globally (i.e. looking at the forest) and locally (i.e. looking at the trees)". This Flexible Information-Processor employs both processing styles either simultaneously, or at least more interchangeably. The discourse of our participants suggests that this process is driven not by the plot or the structure, but rather by the characters, the emotions they elicit and consequences of their "actions" on past, present and future events. Quite often, while developing dialogues and scenes, what has been planned can be abandoned; new ideas are generated, tested and implemented which will have an impact on what has been already written—a classic "domino effect". The choice between new alternatives or ideas leads to the rewriting or suppression of previous dialogues and scenes. These processes occur in a cyclical manner until the task is completed:

> We progress by asking questions. But the choices are made when you write… then you will start all over again and try to pull the thread for a week. Sometimes it leads to an impasse, you realize that it is no longer tenable… Because the scenario is like a game of dominoes: the action of a character on scene three will have consequences not necessarily predictable at the beginning. (S1)

In addition, scriptwriters reported that as one "gets into the skin of the main character", "understands the character" and "makes him or her talk", these are enjoyable moments where intuition, unconscious and automatic processes take over the generation and selection of creative ideas. Empathy with the characters leads to the feeling that they came to life and make the choices instead of the writer, "hijacking or guiding the story line in unexpected ways" as reported by Australian fiction writers (Paton, 2012). This sense of being "carried over" designates a temporary loss of self-consciousness characteristic of a flow state (Csikszentmihalyi, 1996):

> In every story, there is a moment that reminds me of what we feel in an automatic car. There can always be a change of gear decided by the car itself, not your hand or foot. This arrives when the story is on the way (…). There is a sensation of speed and fluidity one can recognise when it happens. (S3)
>
> Sometimes, indeed, a character that one has created or half created, escapes… There is a moment when, as the story gets built, the characters begin to have some rationality in behaviour that can then lead to a new situation, and get the story to progress. (D6)

It is the unconscious that talks. You are about to write something and it takes you in an unforeseen direction. (S1)

Scriptwriters reported on the problems, obstacles, impasses or deadlocks they experience during the writing process: *"the created characters or the situations they are in lead to impasses or deadlocks"* (S2). Solving these problems leads to new developments: *"There are also times when it is laborious. And inspiration happens sometimes because it was laborious, because we made a mistake; because we reached impasses…"* (D20). Writers spend hours "fighting" with the problem, which leads often to fixation or mental block. It becomes impossible to go back, and impossible to consider other solutions: *"There are moments when you feel it just turns around and you are not getting anywhere"* (D6) and:

> Often it is the case that we struggle for a very long time only to settle the problem with one line. I can struggle for two hours for a comma, really! But if I struggle it is because I am not wise enough to notice that the situation is not good. I try to solve the problem artificially, but the characters cannot exist in that situation. (D16)

Most screenplay writers commented on the fact that, during this stage, idea generation and problem-solving occur frequently during unrelated tasks (e.g. emptying the washing machine, bathing a child, taking public transport or working alternatively on another film project in a less demanding phase, etc.). For example:

> Yes (I work simultaneously on different projects). It is difficult and at the same time, I find it good. Because when I struggle and when I am saturated with a project, being able to switch to another project frees my mind and allows me to find answers to the first… (S1)
> Like in painting, you need to let it dry. You write a version and then let it rest… then retake it. Because it's very hard to see it when you have your nose in it. It is exactly like in painting: if you have blue and want it to be red and it is not dry it will become brown. (D16)
> … I'll start by reading it [the proposal of the director], then I'll hang out the washing and maybe there I will have an idea… It could be while giving my daughter a bath… but also when I force myself to write. (S1)

Such environmental influences are not only accidental but often sought by writers. Some of them emphasised their need to write with background noise in cafés or to listen to their favourite music:

> I listen to music, radio… I have always something in my ears when I write. It is like a state of trance. (N5)
>
> I go to coffee shops because there is a soundscape that helps me to be 'absent of myself', distant from the person I am all the time… Being outside, in other circumstances, and in a sort of background noise gives me greater freedom for objects to 'circulate' in my head and around me. I feel that the connections are not similar (when working at home). (S3)

These findings could be partially explained by the mediating role of unconscious thoughts and distractions in facilitating problem-solving and decision-making. McMahon, Sparrow, Chatman, and Riddle (2011) showed that participants who were distracted with easier tasks (listening to music and word search puzzles) made the best decision significantly more often than conscious thinkers and even outperformed participants distracted with more difficult tasks. Finally, Mehta, Zhu, and Cheema (2012) suggested that moderate (70dB) background ambient noise reflecting consumption contexts (e.g. a combination of multi-talker noise in a cafeteria, roadside traffic and distant construction noise) induced processing difficulty which activates abstract cognition and consequently enhances creative performance. Studies on the deliberation-without-attention effect show that a period of unconscious thoughts while making complex decisions can actually lead to better decisions than a period of thorough conscious deliberation (Dijksterhuis, Bos, Nordgren, & Van Baaren, 2006).

The Education of Screenplay Writers

Screenwriting schools ensure that students receive an extensive hands-on experience in how to conceive and write an original feature film, screenplay, documentary or a television series. They achieve this by learning the theory and the practice on how to develop their ideas into a structured script taking into account a given brief, length and format. They also learn the art of pitching their screenplays and collaborating with fellow screenwriting, film-making and acting students.

Our study of both novice students at screenwriting school (FEMIS) and experienced screenplay writers results in a better understanding of the impact of a specific education in scriptwriting on the structuration of the script and the timely completion of the creative work. Our first aim is to examine the stages of the creative work during the learning process as well as individual differences between students. Following this, we will investigate the role of previous education or background of professional writers in the creative process.

Novice Students: Stages of the Creative Process

A group of six students of scriptwriting and film-making studies had eight weeks to create a script starting from a common theme: "A 19 year old women was found dead, murdered by eight knife stabs, in the nave of Notre Dame". The first four weeks were dedicated to collective work and run by a professional scriptwriter to help students produce several alternatives and sketch out a general plan or outline. The last four weeks were devoted to the individual writing of the script. At the end of each week, students had to complete a page of a booklet, which consisted of a self-report part, focusing on the stages of the creative process and the social factors associated with them, as well as an open part in which they were able to record their weekly progress. Based on previous research, 13 stages of the creative process were included: definition of the problem; documentation; consideration of the constraints; insight; association or associative thinking; experimentation, exploration, insight; assessment; structuration; the chance benefit; realisation or implementation; finalisation; and taking a break.

During the first weeks, most students defined the problem with the group and generated several alternatives concerning the overall direction of the story, namely the epoch of the story, the type of film (comedy, thriller...), whether the crime would be the beginning or the end of the story, whether the killer was a man or a woman, what kind of relationship existed between the killer and the victim, etc. They gathered information, read books, watched movies, imagined several stories, started developing character profiles and managing the constraints related to the assignment. Interactions with the group were frequent during this first period. They presented their idea to the group and, with the help of the professional scriptwriter, they had to decide which alternative to choose, sketch out a general plan and structure the story in three or five acts. Although emotions were generally positive over this initial period, a high level of frustration, stress/doubt was observed during the third week. According to some students' comments, they felt frustrated and discouraged by the "obsessive" requirement to make a plan and to structure the script before starting the writing. Indeed, one of the most interesting differences between participants refers to the ease with which they were able to structure and make plans before starting to write the script:

> I have little interest in the formal side of abstract construction: I am no longer confined in this obsessive need of planning and I am happy to finally start writing
> I often have problems with the structure of my scripts, I'm more comfortable in the writing phase
> I can not think a story in an abstract way, I need to immediately shape characters, improvise dialogues

By the fourth week, they started the writing of the dialogues and had to read them in front of the group who offered feedback. After a relative slowdown during the fifth week (school break period), students resumed their writing, editing, changing the structure, rewriting and finalising. For the majority of students, new questions or problems appeared, anxiety and hesitations and blockages were reported in the comments section:

> I'm afraid of losing the thread of the overall plot
> I would like to place flashback but I am afraid it would complicate the story, it becomes incomprehensible
> I do not see how to finish the story in a brilliant way, it's awful
> I am blocked by lack of time, by the crime scene etc.

New ideas and solutions appeared preferentially during this production period. For example, the religious nature of the crime scene, that had been a problem for the students, was solved later in the writing process (week 7 or 8). Students frequently mentioned and selected, in the second period of four weeks, the stages of insight (67%), evaluation (70%), realisation (71%) and finalisation (100%). The parallel increase, over the eight weeks period, of the frequency of insight, realisation and associative thinking confirm that inspiration was high also during the writing or production phase.

Although the development of character profiles took place very early in the process, changes and elaboration were mentioned during the writing phase when the students started to *get into the skin of the main character*. Changes in the script structure (deleting/adding sequences, scenes) occurred continuously as result of new ideas and solutions.

In the last week, some students were dissatisfied or frustrated because of the lack of time to provide a perfect version as the following comments suggest:

> I could have done better in the last ten sequences / imperfect version / this is not a masterpiece / there are a lot of inconsistencies / intrinsic problems to the story / many errors that will irritate the reader

Five students were able to write a screenplay of 100 pages on average (97–110 pages); only one student did not finish within the imposed eight weeks period because he spent lot of time documenting his script and hesitating between alternatives:

> I wasted a lot of time in documentation during the first 5 weeks... I have material for 2 or 3 movies/having too many choices or possibilities is an obstacle... it becomes horribly complicated ... I am irritated by the formal constraints of structuring ... I started to write, but a bit in all directions... I am a embarrassed by my inability to combine everything in a plan that is unified and clear.

Two students showed a high level of stress and frustration during the eight weeks. Interestingly, these students dedicated lot of time to the definition of the problem, documentation, structuring, questioning and management of constraints throughout the entire period whereas the other three students went through these stages in the first three weeks.

Professional Writers: Self-Taught Versus Specific Education

More than half of the professional writers interviewed were self-taught. having a degree in literature, psychology, history, journalism, philosophy or political sciences. Some of them were sarcastic when commenting on how younger generations of French scriptwriters adopted the American way of structuring a story in acts, plot points, etc.

> I do not know all these techniques. I hear more and more about them, with English words of which I know nothing. I think there are a lot of young writers who have been trained in that school and apply with conviction these certainties that after seven and a half minutes, such event must happen and we cannot have a certain type of character without having its negative double, etc. I have always worked empirically. I learned my work with artisans on the workbench. (S13)

The others writers in our sample had graduated in scriptwriting and/ or film-making from New York or Parisian Universities and French schools for film studies (FEMIS, IDHEC, ENSATT). Participants' comments about the benefits of graduating from these schools varied between directors and scriptwriters. According to the directors in our sample, there was no formal training in scriptwriting in their schools or Universities; they considered themselves as self-taught when it comes to writing a script; however, they recognised that they benefited from the opportunity to occupy different positions in the film crew (cameramen, editing…) and learned a lot from making a dozen films over three years. One director commented on the chance offered by the school to "*fail a film without any negative consequences on the opportunity to make another film; to be liberated and not haunted…*" (S4).

Writers who graduated in scriptwriting from specialised schools described the benefits in terms of acquiring the discipline of daily work, a better tolerance to critics, as well as the need to understand the intrinsic motivation or the goal of the story to be written:

I always loved writing, but before attending this school, I wrote sporadically, waiting for 'inspiration'… from the [title of the school], writing became for me a craft, something more concrete… I began to understand the value of daily work. I can no longer work differently: I must sit down almost every day at a table in front of a computer, and force myself to write even if I have no ideas. Then ideas come from the work itself. (S2)

[In this school] I have learned to confront what I wrote to the scrutiny of other – which then allowed me to withstand criticism, re-assessment… the principle to always ask why I wrote that? What effect do I want to convey? These are basic questions, but before, I would not bother to ask them of myself. (S14)

The group of writers with no specific education in scriptwriting stated more often that they were against starting from a preconceived plan; for them, the writing of the narrative is not about "filling in the gaps" of the outline:

I feel I should have written a scene to know how to 'shape' the next one. It's the same for my books, I do not have a plan. (N8)

I am not able to follow the official steps: synopsis, treatment, outline and finally script. I can proceed with the details of some scenes while I still do not know yet what will be the end of the film… I do not start from the skeleton and then add muscles, organs and so on. What I have is some organs, some muscles, a head, arms, maybe bones, and when I have all the parts of the body I try to combine them and make them fit together. (D6)

… This is also why I do not like the outlines and treatments: I find them too predictable. Especially now that they are 'validated' by the producers, so if you change something, they strike back. (S8)

However, they do recognise the use of a basic or "vague" roadmap, of drawings and diagrams made for personal use: "*I have in general a vague idea of the architecture* [of the script], *but a finished plan is something that I never get to have and wouldn't like to make*" (S8).

In addition, experience seems to balance the lack of a rigorous plan or outline. Indeed, the discourse of some writers suggests that some processes as complex as building a story or adopting a 90 minutes format became automatic with time:

I have the impression of having acquired a certain competency regarding these fundamental questions: how to construct a story, depart from one already written, how to write dialogues that don't sound too awkward, how to enter a scene and end it without getting lost. (N8)

Writing is a matter of 'breath'. A story of 1 hour 30, it is a bit like an athletic discipline: the more you practice, the more you are comfortable. But I'm useless when I have to write a short film! What's crazy is that, almost by magic, by experimentation as well, the rhythm of 90 minutes becomes natural. I have great difficulty with other rhythms. (S15)

Writers with degrees in scriptwriting described the benefits of the outline as follows:

- It sets the temporal pattern of the narrative: the plot order, the duration of the story, the rhythm of the story, the shortcuts, acceleration, ellipsis, flashbacks;
- It helps to have an overview of the story, which is one of the major challenges during the writing-rewriting phase;
- It is a useful diagnostic tool; it helps to step back from the manuscript to see what might be wrong in it: "*When I reread the dialogues, I am more in the detail of the scenes. The fact of returning to the outline allows me to take some distance from the script and better see what might be wrong there. Because in the script, there is already the affect: we started to love our scenes, to make our characters speak; we become attached to a particular moment, a particular dialogue… we then struggle to understand where the problem is or to cut when it is too long. The outline, which is much colder object than the script allows it. It is very boring to read*" (S2).

The structuration phase requires a special set of skills, distinct from idea generation or imagining characters. It is a time in which discipline and personal rhythm matter, as well as a logical way of thinking about the storyline:

You can have all the talent and all the literary imagination – which are two essential components of the profession – if you don't have the skills for this tedious task of structuring, you cannot go far. This is something that has more to do with math, a kind of mental structure or consistency: such cause produces such effect. (S15)

The scriptwriters in the second group emphasised that the initial structure of the treatment or the outline is neither static nor a definitive tool. It is most often challenged or updated retroactively while writing scenes and dialogues because of the choices that are continuously made. It acts like a hybrid object containing dialogues. Summarising the differences between different approaches to the initial plan, one of our respondents said:

It is as if you are starting a walk. There are two schools: you either go on an adventure and let yourself be guided by the path, or you carefully study the maps, weather forecast, possible pitfalls, etc. Either way, you can have a very nice walk. (S13)

Conclusion

Our review of social and psychological factors and processes involved in screenplay writing suggests that writing a script is less the result of the creativity of a single writer and much more the outcome of intense interaction among highly skilled, creative individuals. In the film industry, teams evolve from one project to another and collaborate with members with diverse backgrounds, sensibilities, cultural values, aesthetic and stylistic orientations. As in any field of creative production, novel outcomes are the result of a recombination of previous elements which, in turn, relate to the social and cultural background of the participants. There is strong support for the idea that teams with diverse backgrounds and stylistic experiences are able to engage in the "kinds of innovative recombination" that are required to develop a "hit", "that will capture audiences and win critical praise" (Stark, 2009; also Uzzi & Spiro, 2005). The professional writers interviewed for this project worked with various sponsors and were involved in all kinds of productions: author movies, drama, adaptations for TV, comedy or documentary. Their long working experience (an average 24 years) exposed them to diverse styles, personalities and egos which, in turn, lead to the development of tacit knowledge about the work habits of colleagues and furnished them with a rich cultural repertoire. The diversity of backgrounds and experiences of a team, which fosters the emergence of novel ideas, does not guarantee however that these ideas will be recombined and implemented successfully (DeVaan et al., 2014). When working with a co-writer, a sponsor or a director for the first time, our participants reported that the beginning of the creative process is mostly carried out through face-to-face interaction in order to build a certain familiarity and mutual trust. Indeed, a study of the careers of 139,727 individuals who participated 16,507 video games between 1979 and 2009 showed that teams with more dissimilar stylistic experiences outperform teams with more homogenous backgrounds, but only for higher levels of cohesion (DeVaan et al., 2014). The cohesion level is a function of the history of prior collaborations whereas stylistic diversity within the team can be seen as a function of its members' experience with various styles of games: action, adventure, simulation, strategy, sports, racing, etc. The authors argued that

> without some already established, informally codified routines, the dissimilarity of the team's members will sound like only so much noise... Teams with high diversity and high social cohesion are better able to harmonize the noisy cacophony of an (otherwise) excessive plurality of voices, thereby exploiting the potential beneficial effects of cognitive diversity (p.).

As previously reported, the aim of the first meetings with a sponsor is not only to get to know each other but also to develop empathy, to understand the needs and the motivations behind the film project. Empathy contributes to one's ability to understand accurately the thoughts and feelings of others and respond adaptively to others' emotions. Professional writers try to see from the perspective of a director to better express the emotions the latter would like to convey (for more on the relation between perspective-taking and creativity see Glăveanu, 2015). They engage in open and daily discussion to identify the needs, define and redefine current problems. In general, the scriptwriter takes notes and then sketches out a temporary storyboard showing several alternatives, collects, discusses and implements feedback from the sponsor. The development of a detailed characters' background history, personality, psychology, drives, fears, goals, etc., at the very beginning of the collaborative process, is considered by the writers as very conducive to creativity. Writing a script for someone's film seems similar to designing for a user. Indeed, designers create and use fictional users, or personas, they focus on users' needs and produce prototypes which leads to innovative productions that are more adapted to the users. In the work of designers, in fact, several approaches have been developed to encourage precisely this; they include user-centred methods (Norman & Draper, 1986), the "personas" method (Brangier & Bornet, 2010), as well as prototyping approaches, such as IDEO's practice (Brown, 2008).

The present study uncovered a series of general stages of creative activity in the case of expert and novice screenplay writers, as well as different factors that come into play when creating, which were generally convergent between novices and professionals. The process begins typically with a stage of impregnation (preparation for writing), followed by structuration (e.g. writing an extended summary of scenes), then an extensive period of writing and rewriting the actual script and, finally, making final adjustments during the realisation of the film. These phases and their dynamic were mainly reconstructed from interview material and, as such, could be subject to rationalisation and the ordering of process post factum. In this regard, complementing this dataset with observations of actual scriptwriting (recorded through the diary method), was very useful. This exercise pointed to many similarities in terms of the work process. The first week was usually dedicated to defining the problem, gathering documentation, imagining alternatives, etc. In the second and third weeks, most of the students focused on structuring: creating character stories, focusing on key sequences and writing extended summaries. Week four was typically marked by starting to write the script and weeks five to eight were dedicated to writing and rewriting, making revisions and finalising the work. In generalising these findings, we need to take into account though the fact that screenwriting students completed their projects in the context of structured course work.

This dynamic picture of processes and activities in the case of both experienced and novice screenplay writers relates to a common view of scriptwriting as an extended decision-making process, decisions concerning both the content and context of work. According to Nelmes (2007, p. 111), "a crucial part of the scriptwriting process is the ability to make decisions about what works and what does not; there is a continual testing of and discarding of ideas as to their suitability for the story, especially in the early stages and the rewrites".

One of the most important conclusions of our study is concerned with the importance of structuration for a successful and timely completion of the work process. The generation of a "map" before entering the labyrinth of creation is specific for screenplay writers trained in scriptwriting schools and distinguishes them from writers with a non-specific education (history, literature, etc.). For these writers, the role of the sequenced plan is to organise the temporal aspects of the script, give an overview of the story and offer the writer a possibility to step back from the dialogue and consider it in relation to the broader picture. An analysis of student productions points to the same differentiating criterion: generating a detailed plan of the script. The one student who finished the scenario engaged in structuration early on, from the second week of the process. It is worth noting that experienced screenplay writers often do not need to write down a full extended summary of the story and are still guided by the general plan at all times. Their expertise allows them to counterbalance the lack of a rigorous 'map' at the start and finding an effective path within the labyrinth of creation, which is more or less "automatic" and intuitive, an example perhaps of habitual creativity (Glăveanu, 2012). As Nelmes (2007, p. 112) considered, "experience and craft help the writer know how to get characters out of difficult situations, how to get around plot inconsistencies, distract the audience from slips in logic or unbelievable situations, and work through difficult scenes".

William Goldman mentioned that "the writing is never the problem, it is knowing what the structure is and what goes where and who the people are and how this scene relates to this... I have to know all that before I start" (Goldman, 1996, pp. 12–13). This statement can be explained in different ways. Redvall (2009), discussing the role of the first draft, considered it not as a solution to a problem but as a formulation of the problem that allows further developments in writing and rewriting follow-up stages. It could be argued that the sequenced plan has a similar function: it allows the screenwriter to map out the problem space in ways that don't make it a fixed and rigid entity but encourage constant reformulation. For this to happen, the writer needs to evaluate critically the plan and this can make the difference between generating a more or less creative script. Research by Lubart (1994) showed that undergraduate students who were asked to evaluate their written stories earlier in the process (a few minutes after starting) produced more creative outcomes compared to other conditions (including no evaluation). These results were later confirmed by more naturalistic investigations inviting participants to write down their evaluative thoughts during work.

There are certainly limitations associated to our research. To begin, the sample of screenplay writers was quite heterogeneous and, in our exploration, only one main differentiating criterion was observed (previous training). Differences could also be studied, in the future, based on the type of scripts they write (comedies, drama, documentaries, etc.) and their destination (e.g. television and/or cinema). The comparison between expert and novice writers was made difficult by the fact that different data collection methods were used. Future studies could equally try to apply the diary method or another observation technique in the case of professional writers. Moreover, the sample of students was limited as well as the time frame they were offered (eight weeks), something that didn't allow a more prolonged phase of "impregnation" often mentioned by expert screenwriters.

In the end, this kind of research could have important practical implications and orient the work of educators; for example, in our case, highlighting the importance of the impregnation phase and the role of structuration for a successful and timely completion of the script. If "a multitude of paths can lead to a creative story (and an even greater number of paths can lead to a non creative production)" (Lubart, 2009, p. 161), then it is the task of creativity researchers to explore what is specific for each of them as trajectories within the oftentimes intricate but fascinating labyrinth of literary creation.

References

Andreasen, N. C. (2011). A journey into chaos: Creativity and the unconscious. *Mens Sana Monographs, 9*(1), 42–53.

Andreasen, N. C., O'Leary, D. S., Cizadlo, T., Arndt, S., & Rezai, K. (1995). Remembering the past: Two facets of episodic memory explored with positron emission tomography. *American Journal of Psychiatry, 152,* 1576–1585.

Binder, J. R., Frost, J. A., Hammeke, T. A., Bellgowan, P. S. F., Rao, S. M., & Cox, R. W. (1999). Conceptual processing during the conscious resting state: A functional MRI study. *Journal of Cognitive Neuroscience, 11,* 80–95.

Bourgeois-Bougrine, S., Glăveanu, V. P., Botella, M., Guillou, K., De Biasi, P. M., & Lubart, T. (2014). The creativity maze: Exploring creativity in screenplay writing. *Psychology of Aesthetics, Creativity, and the Arts, 8*(4), 384–399.

Brangier, E., & Bornet, C. (2010). Persona: A method to produce representations focused on consumers' needs. In W. Karwowski & M. Soares (Eds.), *Human factors and ergonomics in consumer product design.* Boca Raton: Taylor and Francis.

Brown, T. (2008). Design thinking. *Harvard Business Review* (June), 84–92. https://hbr.org/2008/06/design-thinking.

Buckner, R. L., Andrews-Hanna, J. R., & Schacter, D. L. (2008). The brain's default network: Anatomy, function, and relevance to disease. *Annals of the New York Academy of Sciences, 1124,* 1–38. https://doi.org/10.1196/annals.1440.011.

Buckner, R. L. (2012). The serendipitous discovery of the brain's default network. *NeuroImage, 62*(2012), 1137–1145.

Cole, M. (1996). *Cultural psychology: A once and future discipline.* Cambridge: Belknap Press.

Condor, B. (2010). 'Everybody's a writer': Theorizing screenwriting as creative labour. *Journal of Screenwriting, 1*(1), 27–43.

Csikszentmihalyi, M. (1996). *Creativity: Flow and the psychology of discovery and invention.* New York: HarperCollins.

DeVaan, M., Vedres, B., & Stark, D. (2014). Game changer: The topology of creativity. *American Journal of Sociology, 120,* 1144–1194.

Dijksterhuis, A., Bos, M. W., Nordgren, L. F., & Van Baaren, R. B. (2006). On making the right choice: The deliberation-with-out-attention effect. *Science, 311*(5763), 1005–1007.

Ferguson, B. (2004). *Art, commerce, and values: The relationship between creativity and integrity in the feature film development process* (Order No. 1425324, Pepperdine University). ProQuest Dissertations and Theses,

169p. http://search.proquest.com/docview/305034398?accountid=25340. (305034398).

Ferguson, B. (2014). *Professional creators unveiled: Screenwriters' experiences collaborating in motion picture development teams* (Order No. 3616054, Fielding Graduate University). ProQuest Dissertations and Theses, 235p. http://search.proquest.com/docview/1526012807.

Förster, J., & Dannenberg, L. (2010). GLOMO sys: A systems account of global versus local processing. *Psychological Inquiry: An International Journal for the Advancement of Psychological Theory, 21*(3), 175–197.

Glăveanu, V. P. (2012). Habitual creativity: Revisiting habit, reconceptualizing creativity. *Review of General Psychology, 16*(1), 78–92.

Glăveanu, V. P. (2015). Creativity as a sociocultural act. *Journal of Creative Behavior,* Early View, *49,* 165–180.

Goldman, W. (1996). *Adventures in the screen trade: A personal view of Hollywood and screenwriting.* London: Abacus.

Labro, A. A., & Patrick, V. M. (2009). Psychological distancing: Why happiness helps you see the big picture. *Journal of Consumer Research, 35,* 800–809.

Liberman, N., Trope, Y., McCrea, S. M., & Sherman, S. J. (2007). The effect of level of construal on the temporal distance of activity enactment. *Journal of Experimental Social Psychology, 43,* 143–149.

Lubart, T. (1994). Creativity. In R. J. Sternberg (Ed.), *Thinking and problem solving* (pp. 289–332). New York, NY: Academic Press.

Lubart, T. I. (2009). In search of the writer's creative process. In S. B. Kaufman & J. C. Kaufman (Eds.), *The psychology of creative writing* (pp. 149–165). New York: Cambridge University Press.

Markman, K. D., & Dyczewski, E. A. (2010). Think and act global *and* local: A portrait of the individual as a flexible information-processor. *Psychological Inquiry, 21,* 239–241.

Marková, I. (2003). *Dialogicality and social representations: The dynamics of mind.* Cambridge: Cambridge University Press.

McCrea, S. M., Liberman, N., Trope, Y., & Sherman, S. J. (2008). Construal level and procrastination. *Psychological Science,* 19(12), 1308–1314.

McMahon, K., Sparrow, B., Chatman, L., & Riddle, T. (2011). Driven to distraction: The impact of distracter type on unconscious decision making. *Social Cognition, 29*(6), 683–698.

Mehta, R., Zhu, R. J., Cheema, A. (2012). Is noise always bad? Exploring the effects of ambient noise on creative cognition. *Journal of Consumer Research, 39*(4), 784–799.

Nelmes, J. (2007). Some thoughts on analysing the screenplay, the process of screenplay writing and the balance between craft and creativity. *Journal of Media Practice, 8*(2), 107–113.

Norman, D. A., & Draper, S. W. (1986). *User centered system design: New perspectives on human-computer interaction.* Hillsdale, NJ: Lawrence Erlbaum Associates.

Paton, E. (2012). 'When the book takes over': Creativity, the writing process and flow in Australian fiction writing. *The International Journal of Creativity & Problem Solving, 22*(1). Special issue: Applied creativity and problem-solving in Australia, 61–76.

Pritzer, S. R., & McGarva, D. J. (2009). Characteristics of eminent screenwriters: Who *are* those guys? In S. B. Kaufman & J. C. Kaufman (Eds.), *The psychology of creative writing* (pp. 57–79). Cambridge: Cambridge University Press.

Redvall, E. N. (2009). Scriptwriting as a creative, collaborative learning process of problem finding and problem solving. *MedieKultur, 25*(46), 34–55.

Sawyer, R. K. (2009). Writing as a collaborative act. In S. B. Kaufman & J. C. Kaufman (Eds.), *The psychology of creative writing* (pp. 166–179). Cambridge: Cambridge University Press.

Stark, D. (2009). *The sense of dissonance: Accounts of worth in economic life.* New York, NY and Oxford, UK: Princeton University Press.

Steidle, A., & Werth, L. (2013). Freedom from constraints: Darkness and dim illumination promote creativity. *Journal of Environmental Psychology, 35,* 67–80.

Uzzi, B., & Spiro, J. (2005). Collaboration and creativity: The small world problem. *American Journal of Sociology, 111*(2), 447–504.

6

The Creative Process in Science and Engineering

Giovanni Emanuele Corazza and Sergio Agnoli

The Necessity for Creativity in Science, Engineering, and Society

A discussion of creativity in the domains of science and engineering might appear at first to require qualifying statements. Indeed, it is a somewhat accepted implicit theory that these domains are mainly the realm of rationality, certainly involving clever problem-solving abilities, but in some way distant from the apparently ephemeral world of imagination. Luckily, this implicit theory is actually far from the reality which is experienced everyday by the scientist and the engineer who takes seriously his/her mission to produce advancements in knowledge as well as inventive innovations to improve the conditions of life and

G. E. Corazza (✉)
DEI Department, University of Bologna, Bologna, Italy
e-mail: giovanni.corazza@unibo.it

G. E. Corazza · S. Agnoli
Marconi Institute for Creativity, Pontecchio Marconi, BO, Italy

© The Author(s) 2018
T. Lubart (ed.), *The Creative Process*, Palgrave Studies in Creativity and Culture,
https://doi.org/10.1057/978-1-137-50563-7_6

sustainability of the human species. It is important to underline that it is not sufficient to analyze the characteristics of geniuses, the eminent figures in the history of science and technology who were the elected study cases at the very start of the scientific literature on creativity (Galton, 1869). Rather, this is a discussion that must invest everyone living in today's Information Society (Corazza, Pedone, & Vanelli-Coralli, 2010), as well as those who will experience its medium-long term evolution, which can be identified as the Post-Information Society or the Second Machine Age (Brynjolfsson & McAfee, 2014). In fact, technologies have produced and are producing societal transformations that impact the entire sphere of human activities, from professional to everyday endeavors. The presence of network technologies, that allow everyone to access all databases in the world without appreciable latency has a dramatically important consequence which is unprecedented in history: information is now becoming a commodity. Although it may be difficult today to grasp entirely the full spectrum of consequences of this fact, it can be recognized immediately that if information is a commonly owned asset, then the dignity and self-esteem of human beings cannot rely on pure knowledge of facts. Rather, living a life with significance, making a contribution, requires the transformation of knowledge, i.e., the generation of new ideas, concepts, and products starting from a layer of information shared with all other human and artificial minds. In this scenario, innovation, change, anticipation of the future, and all other activities which can be related to the creative capacities of the human mind assume more and more the essence of tools for survival. Now, the words that William James used to explain the contrast between *rationalism* and *pragmatism* appear to contain nothing less than a prophecy: "*... For rationalism reality is ready-made and complete from all eternity, while for pragmatism it is still in the making...*" (James, 1907, p. 167). This pragmatist view of a world which is brimming with indeterminacy (Shalin, 1986), is a perfectly fitting picture of the present reality, where innovation cycles are reduced to fractions of a year, with more than a tenfold reduction with respect to what was experienced in the Industrial Society. Therefore, it must be realized that rational thinking and clever problem-solving cannot anymore be considered sufficient in science and engineering, as in any other discipline:

creativity is a necessity. Clearly, creativity has both domain general and domain-specific characteristics, and the latter are definitely in the scope of this chapter. But the reader should be warned against the natural tendency of building high walls around specific domains, which can easily turn into stereotypical views and biases: times are mature for true interdisciplinarity, which entails not only cross-pollination between different fields of knowledge, but intrinsic adherence to the fact that transversal understanding of multiple disciplines is a task for everyone. And, in this rapidly evolving society, no discipline is more fundamental and transversal than creative thinking, which should therefore become a basic subject in all education curricula: this requires clearly a scientific approach to creativity. The impact of Information Society itself on the creative process is a topic requiring further investigation (Corazza, 2017; Corazza & Agnoli, 2015a).

The founding ingredient for a scientific approach to creative thinking is to introduce a precise definition for the creativity construct, which is well known to have been actually very difficult to define (Mayer, 1999; Parkhurst, 1999; Rhodes, 1961). However, a standard definition of creativity does exist (Runco & Jaeger, 2012), and can be expressed as: "*Creativity requires both originality and effectiveness*".

As we discussed recently (Corazza, 2016), the standard definition, as well as previous definitions, focuses on the actual achievement of creative success, which requires the recognition by the thinker and/or by external estimators of the originality (that includes novelty, authenticity, and non-obviousness) and effectiveness (that includes value, performance, esthetics, and appropriateness) of the generated outcomes. However, in all domains and in particular in science and engineering, creative outcomes are not always accompanied by public recognition. In fact, there are innumerable examples of the fact that ideas that had full merit were not able to emerge for long periods of time. As an example, at the beginning of the twentieth century, Alfred Wegener proposed first the hypothesis that continents could move, but this idea was rejected as it was considered outrageous, given the knowledge of the time. It was only in the 1960s that the concept of continental drift was accepted as part of the theory of global tectonics (Hallam, 1975). But the fact that creative achievement is not one-to-one with creativity is not only related to the difficulty in getting recognition from the outside world. The fact is that the creative process itself requires in general the exploration of large numbers of alternatives, most of which fail to satisfy the criteria of originality and effectiveness. Creative achievement is a rare gem reserved for those who are persistent enough in their search. For these reasons, we introduced the concept of creative *inconclusiveness* to properly represent the fact that creative activity also includes multiple trials and errors as well as difficulty or altogether absence of recognition. As a consequence, the definition of creativity must subsume both creative achievement and creative inconclusiveness, leading to what we identified as the dynamic definition of creativity: "*Creativity requires potential originality and effectiveness*" (Corazza, 2016). It is the addition of a single word, *potential*, that has the power to change the perspective of our approach to creativity, where growth, dynamism, social interaction, time and space dependence all find a place in our universe of discourse. In particular, how does the concept of *potential* take shape in the domain of science and engineering? It is indeed interesting to note that the dynamic definition of creativity can be related to the work of one of the major philosophers of science, Karl Popper. In fact, it is well known that Popper (1963) theorized that science advances only through a string of so-called "conjectures and refutations". In other words, scientists should propose hypotheses, laws, and theories as conjectures, which have the potential to be held as valid and correct only as long as the absence of contrary evidence is maintained. Therefore it is clear that, according to Popper's view, scientific advancement is related to the ability to imagine or create new hypotheses, the success of which is a dynamic concept, subject to continuous attack and verification. This fits very well with the description of the creativity phenomenon according to its dynamic definition. Similarly, the concept of *paradigm* proposed by Kuhn (1962) as a reference framework for scientific and engineering work, is a dynamic construct that appears to be quasi-static in the periods of so-called normal science, but that it is destined to shift abruptly when a new and better paradigm takes its place. Creative achievement in science and engineering is necessarily limited in time before taking a place in history.

The rest of this chapter is a re-visitation of the famous 4Ps framework by Rhodes (1961) in the light of the dynamic definition of creativity, as applied to the specific domains of science and engineering. We start by discussing traits, motivation, abilities that can be related to enhanced creativity potential for the scientist and the engineer; this is followed by a discussion of the creative process, its principles and methodologies as well as its possible models, which should justify not only achievement but more importantly that vast part of inconclusive efforts that are needed in order to find the rare pearls of success; finally, we address products in the domain of scientific and engineering creativity, which can be classified as hypotheses, laws, theories, discoveries, applications, inventions, as well as the potential of these to be recognized as possessing originality and effectiveness in the real world. Interestingly, considering today's society it can be stated that not only the environment imposes a "press" on the creative thinker, but also the products of engineering creative thinking exert very strong impact on the world, transforming ways of living, necessities and needs of human beings. The domains of science and engineering have much in common as well as many specific characteristics. Our approach is to keep the two domains together as far as possible and then introduce the necessary distinctions.

Traits, Motivation, and Abilities for the Creative Scientist and Engineer

In considering creativity in the scientific domain, and the associated potential for originality and effectiveness, we cannot avoid discussing the main elements characterizing a creator in science and engineering. This person-centered approach to the study of creativity is indeed the cornerstone of creativity research itself. It determined the birth of the scientific study of creativity. As mentioned before, Galton in his book *Hereditary Genius* (1869) collected biographies and described the main intellectual qualities and dispositions of those persons considered to be geniuses, in order to identify the main elements leading to eminence. This approach has been then repeatedly refined, following the progressive specialization of psychological science, and specifically thanks to the generation of specific theories and methods to explore the uniqueness and differences of the individual. Today, the attention is devoted to the specific analysis of personality traits characterizing and distinguishing creators in different knowledge domains, as well as to the analysis of specific cognitive, emotional, and motivational constituents of creative thinking, all the way to the neuroscientific analysis of brain substrates subsuming creative behavior (Mastria, Agnoli, Zanon, Lubart, & Corazza, 2018). Given the dynamic nature of the creativity construct, it is clearly a very complex task to isolate single elements and attribute definitive roles. Therefore, it is necessary to adopt multiple points of view and to look for macroscopic reconciliation. Accordingly, here we follow three complementary approaches to the analysis of individual creativity, each focusing on different elements of the person: personality attitudes, motivation, and the abilities involved in the creative process as applied to science and engineering.

The Creative Personality

Our analysis of the creative personality in the domain of science and engineering attempts to identify the main and distinguishing elements, following one of these two main approaches: the first devoted to the exploration of the personality elements defining creativity in scientists versus non-scientists (e.g., artists); the second exploring those personality elements distinguishing scientists with high creativity potential from those with low potential. Feist (1998) provided a comprehensive meta-analysis of the research on personality and creativity, exhibiting instances of both approaches. In particular, taking into account the Five-Factor Model or Big Five (McCrae & John, 1992), the author showed that the strongest elements in the comparison between scientists and non-scientists emerged in Conscientiousness, Openness, and Extraversion. Specifically, relatively to non-scientists, scientists emerged to be higher on Conscientiousness (which involves items such as: careful, cautious, self-controlled), lower in Openness (which means: conventional, rigid, and socialized), and lower in Extraversion (which means: more deferent, reserved, and dependent). All in all, these characteristics seem to justify several of the common stereotypes about scientists and engineers. However, if we contrast scientists within their own domain, the role of personality traits in determining the potential for creativity becomes clear. In fact, Feist (1998) showed that highly creative scientists are characterized by higher Openness and Extraversion levels than less creative scientists. In particular, they are achieving, ambitious, confident, and have high self-esteem (characteristics involved in Extraversion), as well as curious, esthetic, flexible, and open (characteristics involved in Openness). In other words, the personality gap between scientists and artists tends to vanish in the presence of high creative potential. At the same time, this list of characteristics should not be taken as a cold taxonomy to define the prototypical scientist; rather, it can be used for the purpose of mapping the cognitive, affective, social and motivational elements involved in each specific trait.

Some authors, indeed, formulated hypotheses or developed theoretical models specifying the cognitive elements relating personality traits with creative behavior (Batey & Furnham, 2006; Eysenck, 1993; Kirsch, Lubart, & Houssemand, 2015). This approach is extremely helpful in trying to identify the cognitive mechanisms contributing to the constellation of behaviors organized around a trait. An example of this approach can be found in a recent study that identifies the cognitive mechanism relating the Openness personality trait with creative achievement (Agnoli, Franchin, Rubaltelli, & Corazza, 2015). In this study, we demonstrated that the effect of Openness on creative achievement is mediated by attentional mechanisms, and in particular by the tendency shown by the individual towards processing irrelevant information perceived from the environment. The potential for originality and effectiveness grows if one allows information which is apparently irrelevant into the process, thanks to an open mind, and is then able to use it to produce the outcomes of his/her exploratory exercise in creativity. These results give scientific evidence to the observation by Pasteur that "chance favors only the prepared mind" (1854). The relation between attention and personality in explaining creative performance is indeed important for the understanding of the general creative process, but it is also central for providing new insight in the explanation of the phenomenon of serendipity. Serendipity indeed played a crucial role in many important and revolutionary discoveries in science, such as the discovery of penicillin or X-rays, or in many technological inventions in engineering, such as the microwave oven invention by Percy Spencer (Manley & Laver, 2011). Spencer indeed, while trying to design what would be later identified as a magnetron, stood in front of an active radar and noticed that the chocolate bar he had in his pocket melted. Starting from this aware and open attentional behavior, he investigated the phenomenon and intentionally heated some food with microwaves (starting with popcorn kernels). Further research is needed to consolidate the results on the relation between attention and personality, but the path is open for the identification of the main cognitive and affective elements defining the creative process in the scientific and engineering domain starting from basic personality traits.

Motivational Tendencies Driving the Process

In addition to the analysis of traits, a parallel line of research explores the situational attitudes motivating the creator towards the generation of new ideas. Whereas personality traits remain stable across different situations, the motivation of the thinker does change according to different situations. In fact, motivation can be found in the majority of the theoretical models explaining the creative process (see for example the investment theory by Sternberg & Lubart, 1996). Long debates occurred in the literature about the role of extrinsic and intrinsic motivation for creative achievement, witnessing a historical contraposition between the beneficial effects of internally driven motivation and the detrimental effects of externally driven motivation (Amabile, 1996; Eisenberger, Armeli, & Pretz, 1998; Hennessey, 1989). Nowadays, this contraposition has been tempered by research results demonstrating incremental effects of extrinsic motivation on creativity, in particular when the reward for the task is contingent on creativity (Eisenberger & Rhoades, 2001). But apart from the specific and stable influence of intrinsic and extrinsic drivers (Amabile, Hill, Hennessey, & Tighe, 1994), it can be observed that the motivation of the thinker often resides in the relationship between his/her interests (or characteristics) and the specific focus of the work or problem to be solved (Steiner, 1965). For this reason, motivation has recently been proposed as a mediator between the thinker's personality and the creative task (Agnoli, Runco, Kirsch, & Corazza, 2018; Prabhu, Sutton, & Sauser, 2008).

Considering the specific domain of science and engineering, we can state that the typical objective for the creative process is the search for a solution to a difficult problem, which may be ill-defined and as such defying efforts from many individuals. This constitutional element of the creative process in science and engineering entails a basic feature that clearly distinguishes the scientific and artistic domains. Whereas scientists and engineers typically face problems that essentially derive from a demand of the external world, artists to be authentic should always be driven in good part by internal demand, which can then be solved by an externalized expression that follows either new or accepted artistic canons. Essentially, science and engineering are (mainly) externally driven, whereas art as is (mainly) internally driven. This has important consequences in determining the level of motivation. In the case of the scientific domain, the problem coming from the external world must meet the interests of the thinker, in order to lead to high creative potential. The scientist or engineer must be engaged with the problem to be motivated and believe in his/her capacity to face and solve the problem withstanding periods of creative inconclusiveness. In order to increase the potential contributions, the thinker him/herself should refine the problem definition, to find a sub-problem that best meets his/her interest within a wider open question. As the scientist develops greater expertise within a domain, he/she can indeed search and find specific problems that more positively reinforce his/her interests, i.e., that resonate with his/her intrinsic motivation. Essentially, the problem to be addressed must be charged with the motivational energy of the scientist to provide the necessary perseverance in facing the difficulties during the inconclusive part of the process (Agnoli, Franchin, Rubaltelli, & Corazza, 2018). The mediating role of motivation between perseverance and creativity has been recently demonstrated (Prabhu et al., 2008), and it finds historical confirmation in the work of great scientists of the past. As an example, Keynes (1956) described Isaac Newton's creative ability as the result of an extremely high level of self-efficacy and motivation towards finding new solutions, that led him to think about problems for hours, days, or weeks, until intuition arrived. Equally paradigmatic is the case of Thomas Alva Edison, who declared explicitly that he never considered unsuccessful tentative solutions as failures, but only as chances to learn and to take one more step towards the final solution (Edison, 1948).

Cognitive Abilities Defining Creative Performance

Both personality and motivational elements of scientists and engineers have repercussions on cognitive abilities or creative styles that finally affect creative performance and achievement. Cropley (2015) explored the elements defining the creative process in engineering in particular, describing the essential attitudinal and cognitive styles during the different stages leading to the generation of a new product. He proposed an Innovation Phase Model that, starting from problem recognition and ending with the validation of the product, maps different personality elements and cognitive styles contributing to the inventive process. Specifically, engineers should exploit both convergent and divergent thinking abilities, because both logical analysis and synthesis are needed to reach solutions with potential creativity. However, it can be useful to refine the definition of convergent and divergent thinking styles, not assimilating their meaning to analysis and synthesis. We could more specifically define the convergent thinking style as the tendency of the mind to move towards the single "best" pattern (Runco, 2004), in a typical pattern-matching fashion, whereas the divergent thinking style is intended as the exploratory tendency of the mind to search for all possible alternatives, starting from a common pattern. Guilford (1950) himself believed that divergent and convergent thinking abilities were the main ingredients of creativity. The use of divergent thinking in the scientific creative process is for example evident during the typical problematization phase of reality, that allows the thinker to formulate alternative hypothetical explanations in front of situations that defy immediate explanation. At the same time, the use of the convergent thinking style emerges as a paradigmatic explanation of the ability to find the correct solution to a problem through insight, typical of scientific and engineering creative thinking (Dunbar, 1995). According to Chermahini and Hommel (2010), the convergent thinking style draws on executive functions that keep the thinker on the problem until a valid solution is found. This applies both to routine problems, where the scientist immediately recognizes and applies a method that he/she already knows to find a solution, and to non-routine problems, where the appropriate method is a priori unknown, and a new way must be found, typically requiring a phenomenon of insight (Dow & Mayer, 2004). The importance of the convergent cognitive style in scientific creativity has also recently emerged from the results of the administration of a comprehensive battery of tests to participants from scientific and artistic domains (Agnoli, Corazza, & Runco, 2016), demonstrating that the ability of finding solutions through insight is highly related to basic cognitive abilities (i.e., intelligence), and it is the best predictor for scientific creative achievement (Agnoli, Corazza, Cagnone, & Runco, 2015). Similarly, Botella and Lubart (2015) identify in convergent thinking one of the distinguishing elements of the creative process in scientific invention, that acts in different stages of the inventive process. A clear example of the balance between convergent and divergent thinking in the engineering creative process is the tactic used by the Wright brothers during their work on the first flier: they intentionally argued (Runco & Pritzker, 2011). More specifically, Wilbur and Orville took opposite sides on a specific technical problem, discussed all possible issues in sight (divergent modality), then switched sides and debated again (flexibility), until they reached a final solution and consensus (convergent modality).

Given the importance of convergent and divergent thinking styles for the purposes of creative thinking, it would be of definite interest to understand how these relate to basic cognitive abilities, in order to be able to predict the creative potential of a specific individual. Therefore, more research efforts should be invested to understand and dissect the relationships between convergent and divergent thinking and associative abilities, attentive mechanisms, as well as working memory performance, always keeping in consideration the delicate balance between top-down and bottom-up control and functions. A complementary, but equally important, research route should work towards the clarification of the roles of convergent and divergent thinking modalities in the overall creative process, which we address in the following section.

A Discussion on the Creative Process: Principles and Methodologies

What is the nature of the process that leads to the generation of new ideas in science and engineering? And what are the main domain-specific characteristics that distinguish science from engineering creative processes? One could think that by reading the documents related to major scientific and technological breakthroughs, it could be possible to extract the principles according to which the mind gave birth to the core concepts. Unfortunately, this does not really hold, because the linear way documentation is written does not actually reflect the real process, highly affected by non linearity. Medawar (1991) went so far as to asking "Is the scientific paper a fraud"? We do not believe that there is any fraudulent intention in the way scientific discoveries and inventions are communicated: the fact is that, in general, *proof comes after intuition*. The scientific documents then describe the proof, but in general do not reflect the intuitive leap that is known to have accompanied all major breakthroughs in the fields of science and engineering, which many times involved the introduction of unexpected and sometimes random events, for which there was no a priori justification, but that were essential in reaching the final idea. Therefore, the description of the creative thinking process is a topic of its own, for which the relevant literature is far less than abundant.

Following an approach we recently proposed (Corazza & Agnoli, 2015b), let's assume the simplest possible description of the process by breaking it into three fundamental parts, which we can identify as the 3I: (a) Inception; (b) Ideation; (c) Impact. In particular, Inception includes all the elements contemplating the definition of the focus area, within which new ideas are expected, the activation of the emotional and motivational elements related to that focus area, the gathering and structuring of the information elements which are more or less relevant to that area. The Ideation part contains all of the aware/unaware actions undertaken in order to generate ideas within the focus area starting from the available information, transforming it in non-obvious ways which can use and mix divergent and convergent thinking modalities.

The Ideation element is clearly the home of the potential originality side of the generated concepts. At the same time, Ideation should be always gently guided by self-assessment of the pre-inventive structures, so that the final outcome can also present the characteristics of potential effectiveness, in order to deserve the right to be identified as a creative idea (Silvia, 2008). The Impact part is finally the step in which the creative idea is rendered presentable to the outside world, possibly (but not necessarily) as a solution to the initial problem. The outside world will react to the idea, in a way which depends not only on the intrinsic value of the idea itself, but also on its representation, on the social and cultural environment within which the presentation took place, on the subjective point of view of those (experts or non-experts) who are called upon to express an evaluation. For the same idea, creative achievement or inconclusiveness can happen, intertwine, switch position, depending on the dynamics which are intrinsically embedded into the creative thinking process, in accordance with the dynamic definition of creativity.

It is a simple matter to show how reference models extracted from the creativity literature can be mapped onto this 3I representation. As an example, let's consider the following four models: Wallas (Wallas, 1926), Mumford (Mumford, Medeiros, & Partlow, 2012; Mumford, Mobley, Uhlman, Reiter-Palmon, & Doares, 1991), DIMAI (Corazza & Agnoli, 2013; Corazza, Agnoli, & Martello, 2014), and Geneplore (Finke, Ward, & Smith, 1992). Leaving the details to the interested reader, it can be shown that (a) Inception represents Preparation and initial Incubation stages for the Wallas model; Problem Definition, Information Gathering and Information Organization for the Mumford model; Drive and Information for the DIMAI model; Product Constraints for the Geneplore model. On the other hand, (b) Ideation represents final Incubation and Illumination for the Wallas model; Conceptual Combination and Idea Generation for the Mumford model; Movement for the DIMAI model; Generation of Preinventive Structures for the Geneplore model. Finally, (c) Impact represents Verification for the Wallas model; Idea Evaluation, Implementation Planning and Solution Monitoring for the Mumford model; Assessment and Implementation for the DIMAI model; Exploration and

Interpretation for the Geneplore model. Although it is not in the scope of this chapter to go into the details of this mapping, this brief discussion should be informative enough for the reader to be convinced that a discussion of the creative thinking process based on the 3I representation is more than adequate and significant with respect to the scientific literature on the topic.

What is really crucial here is to discuss the domain specificities of the creative thinking process in view of the 3I representation, as applied to science and engineering. Starting from Inception, which contains definition of the focus and the gathering of relevant information, it is important to dispel the myth that scientists and engineers always take an objective approach to problems. In fact, as Hodson (1986) came to note, scientists are affected by the psychological notion of theory-laden observation, meaning that it is practically impossible to collect and interpret facts without bias or preconceptions. This is because the knowledge that is possessed, albeit at an unaware level, acts as a powerful filter effectively forcing particular interpretations and the elimination of details which are automatically classified as unnecessary or unimportant. This mechanism, which becomes stronger and stronger for increasing levels of a thinker's expertise, is both a blessing and a curse. Clearly, a sharper filter can act to improve the efficiency of the entire process, leading to faster and better selection of focus and information. On the other hand, in many occasions it was a different point of view, an unnoticed detail, an unexpected event, that led to a major scientific or technological breakthrough: in all of these cases, a sharp filter can significantly reduce the creativity potential, which on the contrary requires openness of mind and the ability to process information which is apparently irrelevant, as discussed previously. As an example, let us consider again the discovery of penicillin by Fleming, which is a classical example of accidental discovery requiring acceptance of irrelevant information into the process. Alexander Fleming was a biologist and pharmacologist, who devoted his efforts to exploring the properties of staphylococci; his focus was directed to discovering new antibacterial agents. In 1928 he returned to his laboratory after a summer vacation period and found that a culture of staphylococci was contaminated by a fungus. In fact, before going on holiday he made two mistakes: he didn't sterilize all plates, and he left the laboratory windows open. The quick, normal reaction would have been to simply throw away the contaminated bacteria. This would have been the end of the story. Instead, he opened up to observation: he noticed that in a jelly ring around the fungus, the colonies of bacteria were destroyed, whereas the colonies farther away were not. Then he grew this fungus in culture, and found that it produced a substance (that he named penicillin) that was able to kill numerous bacteria that cause diseases. Two elements were important: not only the accident itself, but also the ability by Fleming to notice, accept, inquire, understand, and finally reproduce a new phenomenon.

Another way to view the limitations in the Inception part of the process is to consider the so-called "allegiance to the paradigm", where the latter is intended according to Kuhn (1970). It is a fact that, once a paradigm is accepted in a scientific or engineering discipline, this forms a research tradition that provides the questions that are worth investigating, dictating what evidence is admissible, and prescribing the tests and techniques that are reasonable (McComas, 1998). Evidently, while the paradigm is very useful in providing direction, it is also a strong boundary that limits the potential for originality. In fact, it happens very frequently that novel ideas are rejected by the experts in a field of knowledge because they do not adhere to the dominating knowledge paradigms.

A final element concerning the Inception part regards the amount of information that is deemed to be necessary to ignite successfully the creative process. It is indeed virtually impossible to discover or invent something new in science and technology without sufficient background knowledge on the major theories, methods, and realizations. But, at the same time, history has shown (Simonton, 1996, 2000; Simonton & Ting, 2010) that great creators were not necessarily extreme experts in their domain. An eminent example of this evidence is given by Guglielmo Marconi, the inventor of the radio. At the time of his first successful experiment, 1895, he was a 21-year old who was definitely living at the margins of the field of electromagnetic theory and applications; he was essentially a self-taught student, with an individual tutor and access to an Italian magazine, "L'Elettricità", thanks to support from his father. However, his apparent unfavorable position and reduced amount of theoretical knowledge allowed him to dare experiment what others could not, being stifled for example by mathematical proofs of the impossibility to transmit radio waves across an ocean. It was only many years after Marconi's successful transatlantic transmission that the ionosphere was discovered, showing once more and dramatically, that *intuition comes before proof.*

Coming now to the second part of the process, (b) Ideation, it is certainly not a simple task to discuss the domain specificities that relate to how ideas are generated in science and engineering, as clearly it is not possible to be exhaustive and we would run the risk to annoy the reader with a long list of examples. Still, it is important to consider at least a few examples from the history of science and technology, to extract what can be considered to be the most important principles that allow the generation of ideas.

First of all, ideation can take place below the level of awareness, as described by the Incubation stage in Wallas model (Wallas, 1926). An important testimonial would certainly be Henry Poincaré (1914/1952), who wrote in detail about his dreamful experiences in idea generation for chemistry and mathematics. However, relying on the unconscious does not give sufficient grounding to a scientific approach to creativity, and certainly is an approach that can hardly be taught in a class. For these reasons, it is necessary to move beyond that and provide more concrete indications. A first element which is important in the generation of ideas for science and engineering is certainly the ability of the human mind to produce and imagine hypotheses. A paradigmatic example in this sense is given by Charles Darwin, who reported in his autobiography that: "*I could not help making hypothesis about everything I saw*" (Barlow, 1958). The effort of generating multiple hypotheses as a vehicle for divergent interpretation is a continuous task for the scientist or engineer who is working on an ill-defined problem. Probing further into Darwin's approach, we can also observe that in formulating a hypothesis he used analogies, borrowing principles from other phenomena.

Notably, in deriving the basic assumptions for the theory of evolution he took inspiration from the principle of "gradualism" (Runco & Pritzker, 2011). Indeed, analogies and metaphors are the strongest tools in the hands of scientists and engineers, as discussed in depth by Gentner (1981). Examples are innumerable: Isaac Newton explained the formation of the moon with the analogy of a ball which is thrown so hard that when it falls it misses the earth and enters into orbit. Claude Berrou, the inventor of Turbo Codes (Berrou & Glavieux, 2003), resolved the problem of designing capacity achieving error-correcting codes, opened by Claude Shannon (1948) some fifty years before, through a daring metaphor: he considered the cascaded decoding algorithms as amplifiers of signal-to-noise ratio. This is a non-obvious analogy, which came to the mind of Claude Berrou only because he was not the greatest expert of information and coding theory, but he was an electronic engineer. From there the idea of using feedback in the cascade (the "turbo" concept) was only natural. Given that the thinker is able to introduce a novel analogy or metaphor, then the necessary creative ability is to be able to find all the possible consequences, certainly using imagination. This was indeed the capacity of great thinkers such as Albert Einstein (Miller, 2011), who used visual thinking skills to imagine experiments in his mind, which led without mathematical proof to the intuition behind the Theory of Relativity. The principles we have mentioned are, of course, but a few: larger sets, that can improve the potential for creativity in science and engineering, can be found in systematic approaches to inventive thinking such as TRIZ (Altshuller, 1984, 1999; Becattini & Cascini, 2015; Ilevbare, 2013). Interestingly, the very complex sets of TRIZ methods can be mapped onto the DIMAI model for the creative thinking process, which turns out to be instrumental for both understanding and application (Agnoli & Corazza, 2013).

The third part of the 3I representation for the creative thinking process, Impact, leads directly into a discussion of the process outcomes, the way they are presented to the outside world, and the reactions that can be expected. In science and engineering, the most important outcomes are in the form of discoveries and inventions. These topics are addressed in the next section.

Creative Products in Science and Engineering: Discoveries and Inventions

In order to guide our discussion on the outcomes of the creative thinking process in science and engineering, let us consider the model of epistemological interaction proposed in Fig. 6.1. This also serves to make a distinction between domain-specific characteristics of science and engineering.

Whereas science is basically concerned with discovering the laws of nature, engineering stems essentially from the necessity of inventing new products for improving life (Corazza & Agnoli, 2015b). Discovery and invention are therefore the main motivational and cognitive drivers in these two macro-domains, as indicated in Fig. 6.1. Considering the science domain, the outcomes of the creative process can take on several forms: hypotheses, theories, and laws, all of which should be both original and effective in order to be considered creative. It is interesting to note that implicit theories exist that map these onto a sort of

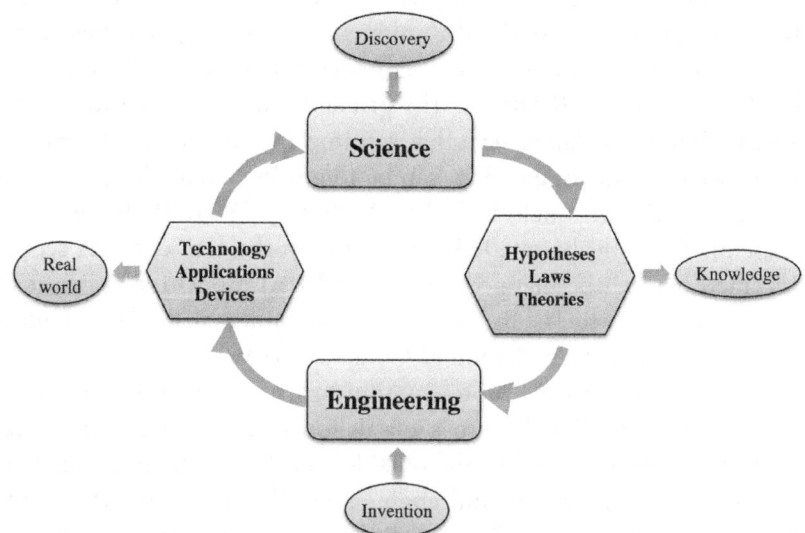

Fig. 6.1 Model for the epistemological interaction between science and engineering

a developmental sequence, with hypotheses becoming theories, and theories becoming laws (McComas, 2002). An implication of this belief is, for example, that hypotheses and theories are less reliable than laws. However, this is a typical bias in reasoning about science, which starts from the erroneous credence that hypotheses, theories, and laws belong to the same epistemological construct. Instead, laws are general principles, describing patterns in nature with sufficient accuracy but without explanation, whereas theories should provide general explanatory frameworks for those general principles (McComas, 2002). A simple example demonstrating the difference between theory and law refers to gravity: we have a law of gravity, but a theory explaining why it operates as it does is still lacking, as of today. Newton indeed accurately explained the relationship of mass and distance to gravitational attraction, but he never speculated about its cause. Now, the creative outcomes of pure science possess the intrinsic value of contributing to knowledge, in its most general sense. This is graphically shown in Fig. 6.1. Scientists, through the derivation of laws and formulation of theories, contribute to the identification of paradigms, which serve as the basis for scientific practice, i.e., for the definition of models from which specific methodological traditions are derived in the scientific research (Kuhn, 1962). As discussed previously, paradigms define the epistemological avenues through which knowledge is developed, in a certain time epoch.

Part of these results from pure science have the potential of being exploited to derive new technology, new applications, through new inventions. In other words, they can be translated from pure to applied science, which is the field of engineering. Indeed, the engineer can be defined as a person who understands scientific facts, finds ways to exploit them, and translates this potential into a practical technological application. In this perspective, we can see that the contributions offered by the engineer relate directly to the real world (see Fig. 6.1), and not to the more abstract realm of knowledge possessed by the human species. The distinction might appear to be very subtle, yet it can be substantiated. Technological applications do not require that they be understood, in order to be useful to the human being. In fact, in today's society people are surrounded by very complex equipment and devices, which are used as black boxes. The more friendly the interface to the user, the shorter and more immediate the learning curve, and the larger the success of the application. A few experts will understand the underlying details, and possibly create a profession based on it, but for the vast majority technology will only represent a tool. The inability to understand the functional principles, especially when the operation fails or differs from our expectation, is a major source of frustration in modern times. From this point of view, it is clear that modern philosophy must be concerned with technology, as was understood for example by John Dewey (Hickman, 1990). Inventions have the potential to impact very significantly on the way we live, either by providing devices that mimic nature or by extending the capabilities of nature itself (Corazza & Agnoli, 2015b). In particular, three main categories of inventions can be envisaged: (1) inventions reproducing nature, which artificially reproduce a natural element of the world, in order to help human beings (e.g., the artificial heart); (2) inventions extending natural capabilities, which introduce devices or services that let people go beyond their natural limitations (e.g., the telescope, the car, the Internet); (3) inventions extending life conditions, which allow life in conditions that are impossible in nature (e.g., the submarine, telepresence, life on other planets). A natural question relates to the perceived necessity for these products of the engineering creative thinking process. It is clear that everyone understands the need to replace parts of the human body with artificial substitutes, if these prove to be safe and efficient, and if they can improve the overall well-being of our species. On the other hand, if an invention introduces a function that was unknown a priori, it seems that this should be classified as a technology-pushed event. The fact is that human beings get immediately used to new devices and functionalities, and what was yesterday an unknown possibility, can easily become tomorrow a commodity, completely taken for granted. A paradigmatic example is certainly represented by the cellular phone, and its latest evolution, i.e., the smartphone. The latter device essentially did not exist before 2005, and in any case, even if did, the telecommunications infrastructure would have been insufficient in order to guarantee the necessary quality of service. At the time of writing, approximately four billion of these devices exist, and are considered totally necessary tools for everyday professional and personal activities. It is an invention with unprecedented impact and rapidity of diffusion. The reasons for this widespread success may be many, but one should be highlighted: the possibility for personalization of a smartphone are nearly endless, depending essentially on the installed applications, available in millions.

To close the circle in Fig. 6.1, we therefore note that because the technological applications produced by creative engineering have such an important impact on the real world, then the very subject of scientific study is affected, and new scientific problems may arise. Again, much evidence can be provided, but let us concentrate for example on the science of human behavior, psychology. Research in psychology today is amply affected by technology, both in terms of instrumentation (e.g., equipment for functional magnetic resonance imaging) and in terms of subject matter (e.g., the impact of the Internet on human relations). Specifically for research on creativity, there are ongoing research endeavors for monitoring and possibly enhancing creative performance in the human brain (e.g., see Agnoli, Zanon, Mastria, Avenanti, & Corazza, 2018; Agnoli et al., 2016).

The epistemological circle between science and engineering therefore closes, only to restart again in a never ending interchange of mutual influences for the benefit of knowledge and well-being of humanity, keeping in mind however that the possible negative effects or misuses of the very same advancements should always be understood, controlled, minimized, and finally overcome with better solutions.

References

Agnoli, S., & Corazza, G. E. (2013). TRIZ as seen through the DIMAI creative thinking model. In *TRIZ Future Conference Proceedings* (pp. 23–33). Paris.

Agnoli, S., Corazza, G. E., & Runco, M. (2016). Estimating creativity with a multiple-measurement approach within scientific and artistic domains. *Creativity Research Journal, 28,* 171–176.

Agnoli, S., Corazza, G. E., Cagnone, S., & Runco, M. (2015). SEM-based analysis of scientific and artistic creative achievement. In *ICIE Conference*, July 1–4. Krakow, Poland.

Agnoli, S., Franchin, L., Rubaltelli, E., & Corazza, G. E. (2015). An eye-tracking analysis of irrelevance processing as moderator of openness and creative performance. *Creativity Research Journal, 27*, 125–132.

Agnoli, S., Franchin, L., Rubaltelli, E., & Corazza, G. E. (2018). The emotionally intelligent use of attention and affective arousal under creative frustration and creative success. *Personality and Individual Differences.* https://doi.org/10.1016/j.paid.2018.04.041.

Agnoli, S., Runco, M. A., Kirsch, C., & Corazza, G. E. (2018). The role of motivation in the prediction of creative achievement inside and outside of school environment. *Thinking Skills and Creativity, 28*, 167–176.

Agnoli, S., Zanon, M., Mastria, S., Avenanti, A., & Corazza, G. E. (2018). Enhancing creative cognition with a rapid right-parietal neurofeedback procedure. *Neuropsychologia.* https://doi.org/10.1016/j.neuropsychologia.2018.02.015.

Altshuller, G. (1984). *Creativity as an exact science.* New York: Gordon and Breach.

Altshuller, G. (1999). *The innovation algorithm: TRIZ, systematic innovation, and technical creativity.* Worchester, MA: Technical Innovation Center.

Amabile, T. M. (1996). *Creativity in context.* New York: Westview.

Amabile, T. M., Hill, K. G., Hennessey, B. A., & Tighe, E. (1994). The work preference inventory: Assessing intrinsic and extrinsic motivational orientations. *Journal of Personality and Social Psychology, 66*, 950–967.

Barlow, N. (1958). *The autobiography of Charles Darwin 1809–1882.* London: Collins.

Batey, M., & Furnham, A. (2006). Creativity, personality and intelligence: A critical review of the scattered literature. *Genetic, Social, and General Psychology Monographs, 132*, 355–429.

Becattini, N., & Cascini, G. (2015). Improving self-efficacy in solving inventive problems with TRIZ. In G. E. Corazza & S. Agnoli (Eds.), *Multidisciplinary contributions to the science of creative thinking* (pp. 195–214). Singapore: Springer.

Berrou, C., & Glavieux, A. (2003). Turbo codes. In John G. Proakis (Ed.), *Encyclopedia of telecommunications.* Hoboken, NJ: Wiley.

Botella, M., & Lubart, T. (2015). Creative processes: Art, design and science. In G. E. Corazza & S. Agnoli (Eds.), *Multidisciplinary contributions to the science of creative thinking* (pp. 53–66). Singapore: Springer.

Brynjolfsson, E., & McAfee, A. (2014). *The second machine age: Work, progress, and prosperity in a time of brilliant technologies.* New York, NY: W. W. Norton & Company.

Chermahini, S. A., & Hommel, B. (2010). The (b) link between creativity and dopamine: Spontaneous eye blink rates predict and dissociate divergent and convergent thinking. *Cognition, 115*(3), 458–465. https://doi.org/10.1016/j.cognition.2010.03.007.

Corazza, G. E. (2016). Potential originality and effectiveness: The dynamic definition of creativity. *Creativity Research Journal, 28,* 258–267.

Corazza, G. E. (2017). Organic creativity for well-being in the post-information society. *Europe's Journal of Psychology, 13,* 599–605.

Corazza, G. E., & Agnoli, S. (2013). *DIMAI: An universal mordel for creative thinking* (Internal Report). Marconi Institute for Creativity.

Corazza, G. E., & Agnoli, S. (2015a). On the impact of ICT over the creative process in humans. In *MCCSIS Conference 2015 Proceedings*. Las Palmas De Gran Canaria.

Corazza, G. E., & Agnoli, S. (2015b). On the path towards the science of creative thinking. In G. E. Corazza & S. Agnoli (Eds.), *Multidisciplinary contributions to the science of creative thinking* (pp. 3–20). Singapore: Springer.

Corazza, G. E., Agnoli, S., & Martello, S. (2014). Counterpoint as a principle of creativity: Extracting divergent modifiers from 'The Art of Fugue' by Johann Sebastian Bach. *Musica Docta, 4,* 93–105.

Corazza, G. E., Pedone, R., & Vanelli-Coralli, A. (2010). Technology as a need: Trends in the evolving information society. *Advances in Electronics and Telecommunications, 1,* 124–132.

Cropley, D. H. (2015). Creativity in engineering. In G. E. Corazza & S. Agnoli (Eds.), *Multidisciplinary contributions to the science of creative thinking* (pp. 155–174). Singapore: Springer.

Dow, G. T., & Mayer, R. E. (2004). Teaching students to solve insight problems: Evidence for domain specificity in creativity training. *Creativity Research Journal, 16,* 389–398.

Dunbar, K. (1995). How scientists really reason: Scientific reasoning in real-world laboratories. In R. Sternberg & J. Davidson (Eds.), *Mechanisms of insight* (pp. 365–395). Cambridge, MA: MIT Press.

Edison, T. A. (1948). *The diary and sundry observations of Thomas Alva Edison.* New York, NY: Philosophical Library.

Eisenberger, R., & Rhoades, L. (2001). Incremental effects of reward on creativity. *Journal of Personality and Social Psychology, 81,* 728–741.

Eisenberger, R., Armeli, S., & Pretz, J. (1998). Can the promise of reward increase creativity? *Journal of Personality and Social Psychology, 74,* 704–714.

Eysenck, H. J. (1993). Creativity and personality: Suggestions for a theory. *Psychological Inquiry, 4,* 147–178.

Feist, G. J. (1998). A meta-analysis of the impact of personality on scientific and artistic creativity. *Personality and Social Psychological Review, 2,* 290–309.

Finke, R. A., Ward, T. B., & Smith, S. M. (1992). *Creative cognition: Theory, research, and applications.* Cambridge, MA: MIT Press.

Galton, F. (1869/1978). *Hereditary genius.* New York: Friedmann.

Gentner, D. (1981). *Are scientific analogies metaphors?* (No. BBN-4604). Cambridge, MA: Bolt Beranek and Newman.

Guilford, J. P. (1950). Creativity. *American Psychologist, 5,* 444–454.

Hallam, A. (1975). Alfred Wegener and the hypothesis of continental drift. *Scientific American, 232,* 88–97.

Hennessey, B. A. (1989). The effect of extrinsic constraints on children's creativity while using a computer. *Creativity Research Journal, 2,* 151–168.

Hickman, L. A. (1990). *John Dewey's pragmatic technology.* Bloomington, IN: Indiana University Press.

Hodson, D. (1986). The nature of scientific observation. *School Science Review, 68,* 17–29.

Ilevbare, I. M., Probert, D., & Phaal, R. (2013). A review of TRIZ, and its benefits and challenges in practice. *Technovation, 33,* 30–37.

James, W. (1907/1955). *Pragmatism.* Cambridge: Harvard University Press.

Keynes, J. M. (1956). Newton, the man. In J. R. Newman (Ed.), *The world of mathematics* (pp. 277–285). New York: Simon & Schuster.

Kirsch, C., Lubart, T., & Houssemand, C. (2015). Creativity in student architects: Multivariate approach. In G. E. Corazza & S. Agnoli (Eds.), *Multidisciplinary contributions to the science of creative thinking* (pp. 175–194). Singapore: Springer.

Kuhn, T. S. (1962/1970/2012). *The structure of scientific revolutions.* Chicago: University of Chicago Press.

Manley, J., & Laver, R. (2011). From the eureka moment to the marketplace. *Policy Options,* 69–70.

Mastria, S., Agnoli, S., Zanon, M., Lubart, T., & Corazza, G. E. (2018). Creative brain, creative mind, creative person. In Z. Kapoula, J. Renoult, E. Volle, & M. Andreatta (Eds.), *Exploring transdisciplinarity in art and science.* Basel, CH: Springer.

Mayer, R. E. (1999). Fifty years of creativity research. In R. J. Sternberg (Ed.), *Handbook of creativity* (pp. 449–460). Cambridge: Cambridge University Press.

McComas, W. F. (1998). The principal elements of the nature of science: Dispelling the myths. In W. F. McComas (Ed.), *The nature of science in science education* (pp. 53–70). Dordrecht, NL: Springer.

McComas, W. F. (2002). *The nature of science in science education: Rationales and strategies.* New York, NY: Kluwer Academic Publishers.

McCrae, R. R., & John, O. P. (1992). An introduction to the five-factor model and its applications. *Journal of Personality, 60,* 175–215.

Medawar, P. (1991). *The threat and the glory.* Oxford, UK: Oxford University Press.

Miller, A. I. (2011). Einstein, Albert. In M. A. Runco & S. Pritzker (Eds.), *Encyclopedia of creativity* (2nd ed.). San Diego, CA: Elsevier.

Mumford, M. D., Medeiros, K. E., & Partlow, P. J. (2012). Creative thinking: Processes, strategies, and knowledge. *Journal of Creative Behavior, 46,* 30–47.

Mumford, M. D., Mobley, M. I., Uhlman, C. E., Reiter-Palmon, R., & Doares, L. M. (1991). Process analytic models of creative capacities. *Creativity Research Journal, 4,* 91–122.

Parkhurst, H. B. (1999). Confusion, lack of consensus, and the definition of creativity as a construct. *Journal of Creative Behavior, 33,* 1–21.

Pasteur, L. (1854, December 7). *Lecture,* University of Lille.

Poincarè, H. (1952). *Science and method* (Francis Maitland, Trans.). London: Dover (Original work published 1914).

Popper, K. (1963). *Conjectures and refutations* (Vol. 7). London: Routledge and Kegan Paul.

Prabhu, V., Sutton, C., & Sauser, W. (2008). Creativity and certain personality traits: Understanding the mediating effect of intrinsic motivation. *Creativity Research, 20,* 53–66.

Rhodes, M. (1961). An analysis of creativity. *Phi Delta Kappan, 42,* 305–310.

Runco, M. A. (2004). Creativity. *Annual Review of Psychology, 55,* 657–687.

Runco, M. A., & Jaeger, G. J. (2012). The standard definition of creativity. *Creativity Research Journal, 24,* 92–96.

Runco, M. A., & Pritzker, S. (2011). *Encyclopedia of creativity* (2nd ed.). San Diego, CA: Elsevier.

Shalin, D. N. (1986). Pragmatism and social interactionism. *American Sociological Review, 51,* 9–29.

Shannon, C. E. (1948). A mathematical theory of communication. *Bell System Technical Journal, 27,* 379–423.

Silvia, P. J. (2008). Discernment and creativity: How well can people identify their most creative ideas. *Psychology of Aesthetics, Creativity, and the Arts, 2,* 139–146.

Simonton, D. K. (1996). Creative expertise: A life-span developmental perspective. In K. A. Ericsson (Ed.), *The road to expert performance: Empirical evidence from the arts and sciences, sports, and games* (pp. 227–253). Mahwah, NJ: Erlbaum.

Simonton, D. K. (2000). Creative development as acquired expertise: Theoretical issues and an empirical test. *Developmental Review, 20*(2), 283–318. https://doi.org/10.1006/drev.1999.0504.

Simonton, D. K., & Ting, S. S. (2010). Creativity in Eastern and Western civilizations: The lessons of historiometry. *Management and Organization Review, 6,* 329–350.

Steiner, G. A. (1965). Introduction. In G. A. Steiner (Ed.), *The creative organization* (pp. 1–24). Chicago: University of Chicago Press.

Sternberg, R. J., & Lubart, T. I. (1996). Investing in creativity. *American Psychologist, 51,* 677–688.

Wallas, G. (1926). *The art of thought.* New York: Harcourt Brace.

7

The Creative Process in Engineering: Teaching Innovation to Engineering Students

Stéphanie Buisine and Samira Bourgeois-Bougrine

To face innovation challenges of the twenty-first century, many companies rely on their engineers to fuel the creative process and set out the roadmap of future technological innovation. Creativity has therefore become a requisite skill for engineers and a part of their basic training. However, there are many ways to implement engineers' creativity and innovation process according to different epistemological approaches. Contrasting philosophies, in particular positivist and constructivist worldviews, determine different design reasoning models and business strategies (Liem, 2014). Positivism refers to a scientific and structured method focusing on identifying the causes influencing outcomes. It is an analytical, problem-centered approach that invests high on the fuzzy front-end of innovation and leads to a waterfall sequential process in which creativity takes centre stage. Herbert Simon's (1973) seminal

S. Buisine (✉)
LINEACT-CESI, Paris, France
e-mail: sbuisine@cesi.fr

S. Bourgeois-Bougrine
Paris Descartes University, Paris, France

© The Author(s) 2018
T. Lubart (ed.), *The Creative Process*, Palgrave Studies in Creativity and Culture,
https://doi.org/10.1057/978-1-137-50563-7_7

research contributed to shape this sequential engineering process based on three major steps: problem setting, (creative) problem-solving, and evaluation of solutions. This view gave rise to many sequential design practices, like the General Design Theory (Tomiyama et al., 2009; Yoshikawa, 1985) and industrial engineering processes organized as a series of stages and gates (Aoussat, Christofol, & Le Coq, 2000; Cooper, 1990; Pahl, Beitz, Feldhusen, & Grote, 2007).

In contrast, constructivism is associated with postmodernism and rejects absolute truth. It considers that reality is a social construct depending on the context: it is a solution-focused approach in which the problem is iteratively co-constructed with the solution (Visser, 2009). This worldview leads to a circular rather than sequential design process, with creative thinking throughout the project. This approach is implemented in many recent design trends, such as information technology (Boehm, 1988), user-centred design (ISO 13407, 1999), agile software development (Beck et al., 2001), design thinking (Cross, 2011), or lean startup (Ries, 2011). Basically, both positivism and constructivism may produce successful outcomes: choosing a process may depend on the project, on the nature of the product to be designed, and most importantly on corporate culture.

Some researchers use an evolutionary metaphor to characterize different business styles and corporate strategies (Picq, 2014): K-type companies are analogous to species that follow a qualitative human-like reproduction strategy (few descendants; high investment in gestation and education; high success rate). In contrast, r-type companies use a quantitative and opportunistic strategy similar to dandelion-like reproduction (many seeds disseminated; low investment; low success rate). The K-type approach may be read as positivist engineering with high investment on the fuzzy front-end, waterfall process, convergence, few new products but high success rate. This kind of strategy can be found in large groups and in the traditional industrial sector. Conversely, the r-type strategy corresponds to constructivist engineering with lower temporal and financial budget, iterative or circular process, divergence, many ideas but high risk of failure, like in startup companies. In natural ecosystems, the K-type strategy tends to outperform the r-type strategy when the competition increases (Picq, 2014). This is why a successful r-type startup company with a constructivist approach may progressively turn its strategy into a K-type positivist approach when growing and gaining investment capacities.

The innovation process is structured by basic methodologies to be selected and arranged in a customized way (sequential, iterative...) for each project. For clarity's sake, we present these methodologies below as a sequential process divided into the four stages of the New Product Design process (Aoussat et al., 2000), knowing that each method can be extracted and used independently or integrated into a constructivist process as well:

- The first stage of the New Product Design process, *Translation of needs*, aims to define functional specifications of the future product to design. This stage involves methods and tools allowing the team to understand better the users, the market, and competitors' products. They include surveys, technological watch, trends analysis, field observations and user studies. Some communication tools exist to share the results of these studies, for example product mappings and inspiration boards to illustrate the state of the art and capture design trends (Bouchard, Christophol, Roussel, & Aoussat, 1999), or Personas to represent archetypes of customer segments (Pruitt & Adlin, 2006). The data collected is finally synthesized through value/function analysis, which results in a list of functional specifications, associated with key performance indicators and target values to be achieved by the future product.
- The second stage, *Interpretation of needs*, draws on the results of the first stage to search for new concepts and new solutions that will meet function specifications and key performance indicators. This is the main creative stage of the process. To conduct it successfully, the engineer's toolbox includes basic creativity techniques such as Brainstorming (Osborn, 1963) and its declinations brain purge, analogies, or problem reversals (Van Gundy, 2005), mind-mapping (Buzan, 1991), etc. Engineers are used to conducting collective creativity sessions in order to maximize divergent thinking through multidisciplinary team and, when possible, integration of users in the session. The creative phase results in a pool of ideas and concepts that are then sorted and ranked using multi-criteria matrices which include the key performance indicators from function specification.

More specific and convergent creativity methods, such as those from the TRIZ framework (Altshuller, 1996; Savransky, 2000), can also be used to model technical/physical problems and find inventive solutions. The second stage ends when a satisfactory concept is selected by the project team to serve as a basis for the new product. In a constructivist process, several different leads from the creative phase(s) might be explored in the project.

- The third stage, *Product definition*, is dedicated to detailed design and materialization of the concept: product architecture, which is sometimes modeled using SADT (Structured Analysis and Design Technique) and/or FAST diagrams (Function Analysis System Technique), choice of technical components and materials, mock-up design, product-process link, Computer-Assisted Design, etc. Intermediate user tests can be conducted on representations of the product concept (3D picture, high- or low-fidelity mock-up, storyboard…). Finally, the product solution, the associated processes and production means can be assessed through FMECA (Failure Mode, Effects, and Critically Analysis).
- The final stage, *Product validation*, aims to validate product design by (1) building an industrially reproducible prototype and (2) having it user-tested. In a constructivist process, the industrially reproducible prototype is not required and user-tests are preferably conducted on low-fidelity mock-up or minimum viable products (Ries, 2011).

The aim of this chapter is to provide insight on what methods engineering students should be trained, how they should be used in the creative process, and the subsequent impact on their creativity in the context of simulated or real innovation projects. We present three pedagogical experiments conducted in three different schools of engineering.

Study 1

The first study took place in a generalist engineering school in Paris (Arts et Métiers ParisTech). The participants were students who were introduced to the above-mentioned New Product Design process

(Aoussat et al., 2000) and the related methodological blocks through a 150-hour class entitled "Product Engineering".

Method

Participants

The sample included 27 students in their final year of engineering studies (4 females, 23 males, $age = 23$ years ± 1). They were rewarded course credits for their participation.

Procedure

The participants engaged in an innovation exercise that they had to perform individually outside of class hours over an 8-week period. To validate the exercise, they had to dedicate 10 working sessions to this project. The goal was to imagine a kitchen for a minivan with the following requirements: Enable cooking, storage of water, dishes, fresh food; be adaptable to most minivans with no modification of the vehicule; occupy no more than 30% of the trunk; be installed in less than 15 min; weight less than 20 kg; comply with security standards, etc.

The participants were provided with a blank booklet to track their process: for each session they had to fill in a self-report of the stage(s) of the creative process addressed (Table 7.1) and an open-ended section to describe the methods used and the intermediate ideas and productions. This methodology of repeated measures was previously tested in research on emotions (Diener, Smith, & Fujita, 1995; Vansteelandt, Mechelen, & Nezlek, 2005; Zelenski & Larsen, 2000).

At the end of the project, the participants were instructed to provide:

- Six different idea sheets corresponding to six kitchen layouts: two for short-term implementation (< 1 year), two for medium-term (between 1 and 10 years) and two for long-term implementation (> 10 years).
- The booklet retracing their process (creative stages, methods used and intermediate productions).

Table 7.1 The 13 stages considered in the booklet

1	Definition of the problem	Focus, explore the theme, the aims, need to create, need to express, challenge
2	Question	Ask, interact with the work, understand
3	Documentation	Capture and search for information, be attentive, always have the project in mind, store information, accumulate, be impregnated, receptive, available, observe, show sensitivity and awareness
4	Consider the constraints	Define constraints, identify a customer's request, set constraints for oneself and define one's rules and freedom
5	Insight	Have an idea, experience the emergence, the sudden appearance of an idea
6	Association, associative thinking	Resonance, play with forms, materials and significations, imagination, daydream, analogy
7	Experimentation, exploration, divergent thinking	Try, modify, manipulate, and test
8	Assessment	Be self-critical, stand back, analyze, reflect, check the quality of a result
9	Convergent thinking, structuration	Crystallize, make a prototype, visualize and structure, establish order, sequences, control and organize
10	Chance benefit	Luck of the environment, aleatory processes, be open to chance events, to take a walk, to accept accidents and chaos
11	Implementation	Transpose, make, illustrate, produce, compose, give shape, apply
12	Finalization, ending	Edit, develop, complete, justify, explain one's work, exhibit
13	Break	Rest, digest an idea, let time pass, do something else

Evaluation of Creative Performance

A multidisciplinary jury of five teachers from the school, all specialized in innovation, evaluated independently the 162 layouts for a functional kitchen produced by the 27 students on a 7-point Likert-type scale (1: not at all creative to 7: very creative). The judges received the layouts to be evaluated in random shuffled order, with no information about the students and no access to booklets. Inter-judge agreement was .80 (Cronbach's alpha), which is very satisfactory.

In addition, in-depth analysis of the 162 kitchen layouts was conducted according to 4 criteria: originality or uniqueness, flexibility or variety, elaboration and integration of technology. An original kitchen concept was unique, surprising, different from the obvious and commonplace. The focus was on the uniqueness of the concept (e.g., proposed by only one student). Flexibility or variety refers to the number of different kitchen concepts proposed (e.g., at least two different concepts among the six designed kitchen). Elaboration measures the amount of detail associated with each kitchen idea. Elaboration has more to do with focusing on each solution/idea and developing it further and adding details. Integrated technologies included green energy, smart or connected kitchens...

The booklets were analyzed as well in order to assess the creative process stages and the methods used by the participants.

Results

Output and Creative Performance

All students managed to produce six layouts for an integrated kitchen. The booklet analysis revealed that the most common aspects considered by students were the reduction of cost and size of the kitchens, the spatial position inside the car, the modalities of use (outside and/or inside the car, while driving), the modularity (functional units as basis of design), practicality (easy to store, deploy and to carry), and technology integration (energy production, water and waste recycling...).

The most creative students came up with original unique concepts of kitchens, different from a classic home kitchen, which could allow new experience for the user such as all-weather kitchen, inflatable or ecological kitchen, remote control food cooking using smartphone, dehydrated food, magnetic levitating modular kitchen, smart or connected kitchen e.g., touch screen, electronic recipes, automated food preparation according the weather and the journey information as well as the available ingredients...

The layouts produced were more or less creative according to the assessment made by jury members. The average jury creativity mark was 4 ± 1.6 with a maximum of 6.2 and a minimum of 2 (1: not at all creative to 7: very creative). To investigate the inter-individual differences, the sample was divided in two groups, C+ and C−, respectively above or below the average (4 ± 1.6). Fourteen students obtained a creativity mark above the average (named C+) and 13 below the average (named C−). The participants were attributed an alphanumeric code according to their rank: S1 for the student with the highest average jury mark ($6.2 \pm .8$) and S27 for the lowest score (2 ± 1.4).

Creative Stages

The booklet completed over the ten sessions revealed the "path" followed by each student to complete the task and solve the problem. The differences between C+ and C− were observed mainly during the last five sessions: C+ were more likely to have "Illuminate, evaluate, associate, experiment and implement" whereas C− continued to "Question, converge, and consider constraints". C+ students used creativity tools up to the very end (seven uses of creative tools in the last session in C+ group vs. 1 use in C− group). The analysis of free comments in the booklets suggested also that some C+ students sought to "summarize" design constraints in some key limitations and seemed to disregard several other constraints. They showed a flexible and even a bold attitude towards the constraints: they did not hesitate to criticize, reinterpret, reformulate, and even circumvent some constraints. In contrast, C− students were continually preoccupied by constraints such as the size of the trunk, the weight and the volume of the kitchen, energy issues, etc. More importantly, they generated new constraints in addition to the initial specifications and tried to find solutions that were feasible within these constraints.

Creativity and Engineering Methods

The analysis of the open self-report part in the booklet, in which the participants recorded their progress, shed light on the development and creativity techniques used during the creative process. A total of 13 tools were applied by students to solve problems and generate ideas, including: individual and collective brainstorming, brain purge, problem reversal, mind mapping, analogies, TRIZ, FMECA, Personas, FAST diagram, SADT, and APTE framework for value and function analysis.

C+ students employed on average 4.2 tools in their process (± 1.6; min $= 2$; max $= 7$) whereas C− used only 2.2 tools on average (± 1.9; min $= 1$; max $= 6$). Personas, mind mapping, brain purge and problem reversal were employed by few students, all of whom were among the most creative (C+). Brain purge is a creativity tool that helps participants empty themselves of their preconceived ideas or any idea they hold dear. This technique was used only once at the beginning of creativity process by a female student (S1) who received the highest creativity mark (6.2 on average). The purge started by a brief documentation on existing kitchens in small flats, boats, camper van etc.; then she wrote down the specifications as well as a drawing of the classical kitchen to avoid absolutely reproducing a classic idea. Indeed, her six alternatives layouts had little in common with the classical kitchen. Four of them were unique, highly original, diverse and included state-of-the-art technologies. It is worth mentioning that S1 applied frequently tools such as mind mapping, brainstorming, personas and analogies during the ten sessions. The quality of execution and output of these techniques showed high standards and allowed the student to experience stimulating divergent thinking. Interestingly, S1 did not use any of the analytical rational techniques such as value/function analysis.

The Persona technique included narratives about different emotional customer experiences and scenarios of use that helped the students to develop some empathy with target customers such as: explorer in Arctic regions or Amazonian forest, nature lovers (ecological kitchen), tradition seekers, elegant and purist design adepts, or technological geeks. Surprisingly, some students did not fully develop the personas; others did not integrate the output of these creative sessions or failed to produce elaborate details into their final layouts. This could be linked probably to an insufficient training or a lack of trust in the benefits of this user-centered design technique.

Value/function analysis was the third most frequently used technique during the early stages of the creative process (the three first sessions). It consisted mainly of reformulating the initial specifications: no new ideas were generated but students felt they gained a better understanding of the problem and declared they were ready to get started. However, among the 11 students who applied function analysis, 8 (73%) did not come up with any unique or original idea. FAST diagrams, which display functions in a logical sequence and prioritize them, were used by only one student (S25) during the 4th, 5th, and 6th sessions. The creative performance of this student was in the bottom three (average creativity mark: 2.2 ± 1.3) despite a very structured approach and the use of a total of 6 engineering methods.

Discussion

This study provides several insights on the relation between engineering methods, how/when they are used in the process and the subsequent creative performance. The main features of the creative process of C+ students were: their reduced attention to constraints, their use of creative/diverging tools until the very end of the project (constructivist process), and their reflexion on user needs through personas.

We observed that students who both alleviated technical constraints and adopted users' viewpoints produced the most creative outcomes. This result should be read in conjunction with recent analyses of corporate innovation strategies worldwide, in particular the 2014 study

Global Innovation 1000 (Jaruzelski, Staack, & Goehle, 2014). It shows that three basic strategies can be found in innovative companies: Technology-driver (whose priority is to develop products of superior technological value), Market-reader (which focuses on creating value through incremental innovation and customization of products), and Need-seeker (which aims to find unstated customer needs of the future, and to be the first to address them). Although the three strategies all possess their own success stories, a long-term analysis shows clearly that Need-seeker outperforms the two other strategies in terms of financial return on investment (Jaruzelski et al., 2014). In line with this global trend, our results suggest that focusing more on users and less on technical constraints leads to more creativity, and that engineering students should be trained to do so.

This way of managing innovation projects is not obvious, particularly in France. Indeed, the Technology-driver model remains dominant in France (60% of innovative companies; Péladeau, Romac, Rozen, & Sevin, 2013) and Need-seeker model struggles to emerge (17%). In contrast, Silicon Valley firms are the most likely to follow a Need-seeker model in the world (46%). Innovation analysts recommend therefore developing the Need-seeker strategy in France in order to stimulate innovation and thereby economic growth (Péladeau et al., 2013). In this respect, the Persona method is a convenient, low-cost approach to support engineers' empathy with users, but the benefits might be even stronger if engineers were used to integrating real users in the innovation process, through e.g., interviews, field observations, or user tests. Our sample students were actually taught these methods in the Product Engineering class, but they did not use them in this project.

We drew on this set of results to build a new training program for teaching innovation to engineering students, with the following characteristics:

- We decided not to integrate value/function analysis in the innovation process, in order to avoid too much focus on technical constraints and on evaluation criteria.
- We integrated field and user studies as mandatory steps, with dedicated sessions planned in the program.

- We designed the program in order to foster a constructivist process including several rounds of analysis, creativity and design throughout the pedagogical project.
- As corporate innovation projects are always conducted in teams, we decided also to make students work in groups rather than individually.

Study 2 reports on the implementation of this second training program.

Study 2

This study took place in another general engineering school in Paris (Ecole d'Ingénieurs du CESI), which is known for its innovative pedagogy: this school is class-free and learning relies exclusively on active pedagogy through projects. The participants were students who had chosen the "Innovation" specialty for their final year and engaged in a 210-hour innovation program including 175 hours of group project and 35 hours of personal work.

Method

Participants

The sample included 30 students in their final year of engineering studies (9 females, 21 males, age $= 24.3$ years ± 1.6). Their participation validated partly a semester of their engineering curriculum.

Procedure

Five groups of six students were composed on the basis of an initial deliverable in which the students had to describe their motivation for the Innovation program and list examples of products they would like to study, to improve or to create. The groups were composed by the experimenter and attributed five different projects (one project for each group) in accordance with students' interests. The sample projects included two assignments provided by partner companies, two entrepreneurial projects provided by students, and a fictitious project provided by the experimenter on the basis of the group members' interests. The projects were focused on different products (three goods, two services) and therefore had different specific goals but they all consisted of starting with a concept and making it become a concrete reality at the end of the project. This required refining the response to users needs and expectations, refining the concept, positioning strategically the product with comparison to existing ones on the market, elaborating a detailed design, and developing a business plan.

The groups had 5 full-time weeks (i.e., 175 hours) to achieve their project. They were guided through an innovation process (Table 7.2), had to produce daily deliverables and were provided with mentorship from several experts.

Table 7.2   The methodological steps imposed along the five weeks

Weeks	Methodological steps
1	Technology watch
	Use analysis (field study)
	Creativity
2	Creativity
	Materialization of ideas
	Mentoring committee
3	Materialization of ideas
	User tests (field study)
	Creativity
	Patent watch
4	Creativity
	Mentoring committee
	Marketing
	Business plan
5	Intellectual property
	Creativity
	Business plan
	Mentoring committee
	User tests (field study)

Feedback on Creative Performance

Twelve experts from the school and from partner institutions participated in the mentoring committee that met three times during the project. The experts represented different specialties such as technological
innovation, user-centered innovation, industrial design, finances, and
strategy. The experts gave qualitative feedback on each project and deliv-
ered customized advice to each group.

Students were also invited to give individual feedback on their experience as apprentice innovators. At the end of the project, they had to
self-assess the contribution of each methodological step on their creative performance (on Likert-type scale from 1: not important at all to 7:
very important), and to indicate what had been most striking to them
in this pedagogical project (open-ended question).

Results

The five projects were very different from one another but all groups
managed to gain one or several supporters within the experts' committee. For example, one group proved very flexible in the solutions
imagined and also achieved a high degree of elaboration. Another group
produced a very original business model. A third group combined existing technologies to provide a new solution to unmet societal need. All
groups' productions were acknowledged as creative and attested by a
Soleau envelope (proof of invention) submitted to the French National
Industrial Property Institute. Moreover, one of the projects resulted in a
patent application, currently in progress, and another one resulted in a
startup creation. Experts' opinions were nonetheless much contrasted,
some of them being more receptive to technological innovation, some
to business plans, and some to response to user needs.

Regarding students' opinions of which methods were pivotal to their
creative performance, the results show that training on creativity methods was ranked first, then mentorship, group composition, and user
studies (Fig. 7.1).

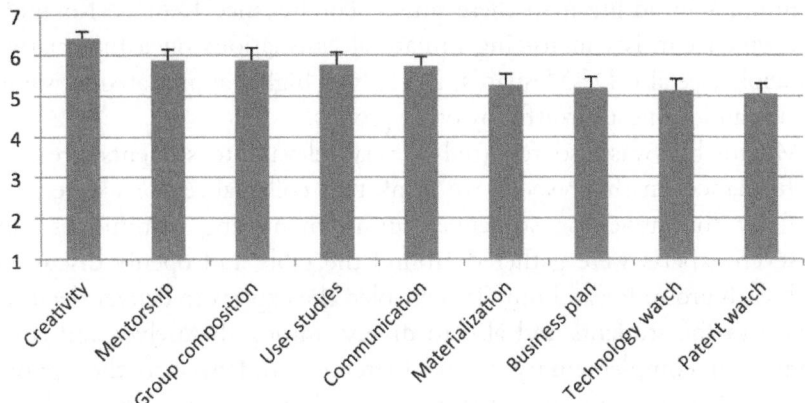

Fig. 7.1 Subjective evaluation of the impact of methodological steps on students' creativity. Likert-type scale from 1: not important at all to 7: very important

Discussion

Our results did not enable us to distinguish between the five projects in terms of creative performance, because each project had its own specificities, strengths and weaknesses, and got support from at least one expert from the mentoring committee. The overall constructivist process seemed natural to students and none of them reported any redundancy between e.g., the five iterative creative steps organized throughout the project. Some of them even suggested that the business plan, which was introduced in the fourth week, should have been initiated from the very beginning of the project.

Students judged creativity methods as central to their performance and 21 students out of 30 cited them as one of the most striking learning outcomes of the program. They were actually introduced to basics of the following methods: brainstorming, mind-mapping, visual projection, Kent & Rosanoff list, problem reversal, analogy, trends, and discovery matrix. Informal comments collected during the project suggested that brainstorming and problem reversals were the most widely used techniques and that each group also had its own favourite techniques in addition to these musts. For instance Kent & Rosanoff list, which consists in forcing unnatural associations on a bissociative principle (see also DesMesnards, 2011), was highly appreciated by some groups and found unfruitful by other groups.

Mentorship was also reported as very relevant to students' creative performance. In this respect, we think that collegial sessions were particularly formative for students: during mentoring committees, up to seven experts were gathered around the table and openly discussed with each group for an hour. This enabled the experts to deliver detailed advice to the students and also to discuss among themselves and confront their complementary views. Therefore, students had the opportunity to understand that there was neither a unique approach to innovation nor straightforward answers to their doubts and questions.

Group composition obtained the same importance score as mentorship to account for creative performance. In the open-ended section of the questionnaire, many students commented on this effect with highly positive terms, explaining that they had experienced powerful group cohesion during the project. It was a striking experience for 13 students out of 30. Some of them underlined that they had learned to work better in a group and take advantage of their differences. Although students in this school were already used to working in groups, they had never experienced such a long (five weeks, full time) collaboration. We may point out that both group composition and project assignment were imposed by the experimenter on the basis of a one-page deliverable produced by the students. Although unusual and risky, this procedure seemed effective, as attested by students' evaluations and by further indirect evidence: on the first day (project launch) when we circulated group composition and project assignment to the students, we allowed one permutation by group, but observed that only 3 students out of 30 actually changed groups. Likewise, after two weeks and a half of group work (mid-term project review), we offered a new opportunity to change groups and once again only 3 of them decided to change groups, which suggests that the majority of students were satisfied with their group.

Finally, user studies were the fourth method acknowledged as important for creativity. In this respect, we achieved our goal of promoting user integration in engineers' innovation process. Three days of the training program were dedicated to field studies and students had a special authorization to "get out of the building" on those days. They could observe uses, meet potential end users of their products and interview them, get feedback on their concepts and better capture user needs and expectations. However, we would describe the results obtained as a Market-reader rather than a Need-seeker process. Students could indeed greatly improve and refine their concepts, but they did not generate disruptive ideas from user studies (or, at most, only one group did). In contrast, the Need-seeker approach is assumed to turn into radical innovation, make future needs arise and generate undreamed of concepts. Examples of companies known for their Need-seeker strategy include Apple, Tesla or Procter & Gamble (Jaruzelski et al., 2014)—we suspect that our students did not live up to these prestigious references. This is why we decided to further improve the previous innovation process and guide students through a stronger Need-seeker-like approach.

Need-seeking represents today the pinnacle of innovation and is often attributed to geniuses or visionaries; hence there are relatively few known methods for structuring it. The Lead-User method (Franke, Von Hippel, & Schreier, 2006; Von Hippel, 2005) may be the most effective one to date: by definition, lead users are precursors, and are at the leading edge of important trends in the market. Involving lead users in an innovation project may grant access to needs that will later be experienced by many users and therefore may open successful innovation opportunities, as in companies like 3M. However, this method seems hardly applicable in a five-week pedagogical project because lead users are difficult to find and may require up to several months to be identified (Von Hippel, 2005) before being eventually integrated into the innovation process. In contrast, the low-cost Persona method used by some participants in Study 1 enabled them to generate creative ideas because it involved "extreme" (although fictitious) users, for example explorers in Arctic regions or the Amazonian forest. We decided to elaborate an intermediate Need-seeker method between the Lead-User and Personas that would involve "extreme" users, although not as unique as lead users and not as fictitious as Personas. Study 3 reports on the testing of this original method with a new population of engineering students.

Study 3

This study took place in an engineering school specializing in biology and biotechnology oriented towards the pharmaceutical, cosmetic, food and environmental industries (Ecole de Biologie Industrielle). The participants were students who had chosen the "Engineering design" option for their final year. The present study was conducted as part of a nine-hour "User-Centered Innovation" class in this option. Given its limited timeframe, the pedagogical project focused on the Need-seeker step only and did not address the whole innovation process.

Method

Participants

The sample included 55 students in their final year of engineering studies (50 females, 5 males, age = 23 years ± 1). Their participation contributed to the validation of their option.

Procedure

Students composed 6 groups of 9–10 members and each group chose an existing product as the starting point of its project. Most of the students in this school are experts in and passionate about cosmetics and strive to pursue careers in this industry. Therefore most of the groups chose a cosmetic product (e.g., nail polish, eye liner, powder foundation), which offered incidentally a very nice challenge to this experiment. The goal of the project turned into finding out new unmet needs related to existing products from a hyper-competitive market with intensive innovation activity.

The Need-seeker method elaborated for this project was inspired by Universal Design (Buisine, Plos, & Aoussat, 2011; Vanderheiden, 1997; Vanderheiden & Tobias, 2000). In many aspects, universal design meets usability principles (ISO 9241-210, 2010; Nielsen, 1993) but generalizes the approach to *all* users (be they young, old, disabled, tall, small...) and not only to target users of a given product (sometimes corresponding to very narrow market segments). In line with this principle, our method named "Off-target user" consists mainly in testing a product outside of the target user population. We hypothesized that focusing on users with special needs would feed the much-vaunted Need-seeker strategy by renewing the view we take on a product, revealing latent needs that are not expressed by target users, and highlighting new original needs. For example, if we study children needs (e.g., beginner readers, narrower vocabulary, shorter stature, weaker force...) while designing a product for adults, this may result in a more intuitive product, with higher usability for adults, elderly people, disabled, foreigners who speak less well the language, etc. The same reasoning applies to senior needs (i.e., viewing and hearing disorders, lower dexterity, memory disorders, etc.), which are likely to help us design more intuitive products for able-bodied users.

The pedagogical projects were therefore aimed to identify unmet, latent or unknown needs related to the products of interest. The groups had to conduct user tests with five target users and five off-target users, confront the needs identified in the two conditions and select an innovation challenge for this product for the next ten years.

The course of the project was designed as follows: students attended a four-hour class introducing them to user-centered innovation, Need-seeker strategy and finally the original Off-target-user method to implement. Then they had two hours for (1) composing the groups, (2) choosing their product of interest, and (3) setting out their protocol for target and off-target user tests. They had subsequently two weeks to conduct the tests outside of class hours, analyze and synthesize the data. The final three-hour class was dedicated to project defense.

Results

Instead of reporting each group's findings, we describe in this section the detailed results of the group that worked on nail polish. We chose this group because it exceeded the initial instructions and conducted a more complete need-seeking process, with a brain purge creativity session, technology watch, and market research in addition to the methods required in the exercise. For this reason, their project gives a wider picture of the contribution of off-target user testing for need seeking. Also, this group published its study (Mear, Moreau, Moussour, Moussour, & Buisine, 2015).

The initial brain purge was conducted with group members only, which was a very homogeneous group of 10 women, aged 21–24 years, with the same training background, all nail polish users (and some of them expert users). The brain purge was dedicated to finding ideas for improving existing products. The main improvement avenues that were identified were e.g., avoiding formula drips, improving application accuracy and reducing drying time. Technology watch enabled them to find original application techniques—including nail art techniques—as well as innovations in the formula (extra-fast dry, thermo-responsive, anti-aging, long-lasting, nail foundation, etc.). Market research confirmed that the domain was very dynamic, with sales in constant growth since 2006 and more than 10 million bottles sold each year. The group also conducted a survey on a sample of 23 women aged 10–74 years indicating that the first nail polish application occurs at 18 years old on average (6–30 years old) and may continue throughout lifetime.

Target user tests were conducted with 5 women aged 18–68 years, who were expert to casual users of nail polish. Off-target user tests were conducted with 2 children and 3 men aged 4–56 years, non-users of nail polish. They were invited to paint their fingernails of the two hands and think aloud throughout the task. Afterwards they had to perform an auto-confrontation (Mollo & Falzon, 2004) and provide further comments on their nail polish experience while watching the video recording of their activity. They were finally interviewed about avenues and/or suggestions for improvement of nail polish products.

Most of needs reported by target users concerned the formula (viscosity, dry time, smell, easiness to remove). They complained generally about the long time required to paint fingernails. They did not comment much on the devices, just mentioned that the brush used for the test was not flexible enough and too small. On the contrary, off-target users made a lot of comments on the devices: bottle plug difficult to screw and unscrew (in particular with fingernails freshly painted), brush difficult to handle (in particular with fingernails freshly painted), bottle difficult to hold, etc. Also they mentioned the difficulty to paint their fingernails of the dominant hand (with their non-dominant hand) and to paint the thumb because its orientation is different from the other fingers. These needs are so obvious that target users did not mention them. We think that these are nonetheless actual needs, and may improve target users' experience if they were met. Indeed, target users interviewed in this study were still 60% dissatisfied and 80% found nail polish application difficult (this reached 100% of off-target users).

Discussion

This study suggests that testing a product with off-target users could be a smart way of highlighting basic and unmet needs as the starting point of a Need-seeker innovation project. The other groups participating in this study obtained similar results with different products (two other cosmetic products, but also two types of food packaging and a hair straightener). Finding off-target users for cosmetic products and the hair straightener was particularly easy because the students could involve men in their sample. For food packaging all human beings are potential target users, but in this case the students involved "extreme", or non-standard users: they conducted their tests with children and elderly users, with the same effectiveness in identifying unmet needs with comparison to middle-aged users. Other valuable extreme users could be found in people with perceptive, motor or cognitive impairments, but they may be more difficult to find in such short pedagogical projects.

The main advantage of involving off-target or extreme users was to highlight unmet although obvious needs. Leading engineers to (re-)discover them is likely to stimulate their creativity and result in new, original and hopefully more usable devices. We speculate that this could be the creative process that was followed in information technology to achieve the highly usable devices we have today: questioning and re-examining the fundamentals of interaction to create more usable interfaces. Famous companies like Apple have built their reputation on this kind of achievements despite sometimes lower technological capacities of their product with regard to their competitors.

Data collected by the groups in this study suggested also that off-target and extreme users showed less cognitive fixations on existing products and generated more divergent (uncensored, fanciful, ambitious) ideas to improve existing products. However, the timeframe of this pedagogical project did not enable the students to use the study outcomes and engage properly in a constructivist creative process. This could be the aim of a future experiment.

In any case, we consider that our goal was met to provide students with a simple method likely to support a Need-seeker innovation strategy, an approach that is currently insufficiently developed in many companies (Péladeau et al., 2013). Moreover, according to informal comments of the students, Off-target-user method enabled them to see the product through users' eyes instead of engineers' eyes, which is an achievement in itself.

General Conclusion

In this chapter we reported on three pedagogical experiments related to teaching innovation to engineering students. The first study was an attempt to analyze systematically the relations between reasoning processes, engineering tools, and creativity. For this purpose we elaborated quite an artificial situation, with a single project addressed in parallel by 27 students, using individual procedures and many traceability constraints (imposed number of sessions, self-reports, booklet to complete, etc.). This was the price to be paid for gaining reliable insights and an understanding of how to design effective pedagogical programs.

The second study was partly designed on the basis of these insights and implemented more realistic situations, with real projects conducted in groups, during working hours—a situation analogous to what students might experience in their (future) professional life. The methodological counterpart was that the five projects turned out to be impossible to compare in terms of creative performance. This study was nonetheless informative as to how the process and the methods were experienced by students, and evidenced further limitations about how they take advantage of user studies to innovate.

The third study enabled us to beta-test the Off-target-user method, which is usable by students and likely to help them to see what Need-seeking is like. The results were very encouraging and call for further study: the method now has to be integrated into a full-length innovation process in order to assess its impact on creativity and innovation.

The "best-of" pedagogical innovation process drawing on this set of results would have the following characteristics:

- A full-constructivist process concerning all dimensions throughout the project (analysis, creativity, evaluation, business plan...)—the sequential waterfall process may become relevant when students get experienced;
- An initial Need-seeker approach fed with Off-target/extreme users and/or Personas and/or Lead users;
- Field studies as mandatory steps;
- Not too much focus on constraints and function analysis—although important for routine design, they may be counterproductive in innovation;
- Mentorship, for example in the form of collegial sessions;
- Group work, and ideally *multidisciplinary* group work—this is a major limitation of the pedagogical experiments presented in this chapter to have entrusted innovation projects to too homogeneous groups in which engineers were among themselves.

In addition to promoting multidisciplinarity, we believe that two main directions should be investigated to further leverage engineers' creativity: the first one consists of developing new creativity methods and tools (see e.g., Afonso Jaco, Buisine, Barré, Aoussat, & Vernier, 2014; Guegan et al., 2015; Schmitt, Buisine, Chaboissier, Aoussat, & Vernier, 2012); the second one relies on orienting engineers' creativity in relevant and original directions through prospective methods (e.g., Barré, Buisine, & Aoussat, 2014; Barré, Buisine, Guegan, & Aoussat, 2014; Nelson, Buisine, & Aoussat, 2013; Nelson, Buisine, Aoussat, & Gazo, 2014), and particularly Need-seeker ones, as exemplified in this chapter.

Acknowledgements Study 1 was partly funded by ANR-CREAPRO project. We are very grateful to the experts who participated in study 1 and study 2 and of course to the 112 engineering students from 3 different schools who composed our experimental population.

References

Afonso Jaco, A., Buisine, S., Barré, J., Aoussat, A., & Vernier, F. (2014). Trains of thought on the tabletop: Visualizing association of ideas improves creativity. *Personal and Ubiquitous Computing, 18,* 1159–1167.

Altshuller, G. S. (1996). *And suddenly the inventor appeared.* Worcester, UK: Technical Innovation Center.

Aoussat, A., Christofol, H., & Le Coq, M. (2000). The new product design—A transverse approach. *Journal of Engineering Design, 11,* 399–417.

Barré, J., Buisine, S., & Aoussat, A. (2014). PLT (Persona Logical Thinking): A method to generate user requirements for multidisciplinary design teams. In *AHFE 2014 International Conference on Applied Human Factors and Ergonomics.*

Barré, J., Buisine, S., Guegan, J., & Aoussat, A. (2014). Le caractère ludique comme levier de performance pour l'anticipation des besoins des utilisateurs. In *ErgoIA 2014 Colloque francophone sur l'Ergonomie et l'Informatique Avancée.*

Beck, K., Beedle, M., van Bennekum, A., Cockburn, A., Cunningham, W., Fowler, M., … Thomas, D. (2001). *Manifesto for agile software development.* agilemanifesto.org.

Boehm, B. W. (1988). A spiral model of software development and enhancement. *IEEE Computer, 21,* 61–72.

Bouchard, C., Christophol, H., Roussel, B., & Aoussat, A. (1999). Anticipation and integration of trends in design and engineering design. In *Proceedings of International Conference of Engineering Design (ICED).*

Buisine, S., Plos, O., & Aoussat, A. (2011). La conception universelle, inclusive (fiche pratique). In *Déployer l'innovation.* Paris: Techniques de l'Ingénieur.

Buzan, T. (1991). *The mind map book.* New York: Penguin Books.

Cooper, R. G. (1990). Stage-gate systems: A new tool for managing new products. *Business Horizons, 33,* 44–54.

Cross, N. (2011). *Design thinking: Understanding how designers think and work.* Oxford: Berg/Bloomsbury.

DesMesnards, P. H. (2011). Mettre en œuvre la technique des rencontres inattendues: La bisociation. In *Déployer l'innovation*. Paris: Techniques de l'Ingénieur.

Diener, E., Smith, H., & Fujita, F. (1995). The personality structure of affect. *Journal of Personality and Social Psychology, 69,* 130–141.

Franke, N., Von Hippel, E., & Schreier, M. (2006). Finding commercially attractive user innovations: A test of lead-user theory. *Journal of Product Innovation Management, 23,* 301–315.

Guegan, J., Maranzana, N., Barré, J., Mantelet, F., Segonds, F., & Buisine, S. (2015). Design and evaluation of inventive avatars for creativity and innovation. In *ICDC 2015 International Conference on Design Creativity*.

ISO13407. (1999). *Human-centred design processes for interactive systems.* Genève: International Organization for Standardization.

ISO9241-210. (2010). *Ergonomie de l'interaction homme-système – Partie 210: Conception centrée sur l'opérateur humain pour les systèmes interactifs.* Geneva: ISO International Organization for Standardization.

Jaruzelski, B., Staack, V., & Goehle, B. (2014). Proven paths to innovation success. *Strategy+Business, 77,* 2–16.

Liem, A. (2014). Toward prospective reasoning in design: An essay on relationships among designers' reasoning, business strategies, and innovation. *Le Travail Humain, 77,* 91–102.

Mear, S., Moreau, S., Moussour, M., Pensé-Lhéritier, A. M., & Buisine, S. (2015). Application de méthodes d'innovation centrée utilisateurs au secteur cosmétique. In *Confere 2015 Colloque francophone sur les sciences de la conception et de l'innovation*.

Mollo, V., & Falzon, P. (2004). Auto- and allo-confrontation as tools for reflective activities. *Applied Ergonomics, 35,* 531–540.

Nelson, J., Buisine, S., & Aoussat, A. (2013). Anticipating the use of future things: Towards a framework for prospective use analysis in innovation design projects. *Applied Ergonomics, 44,* 948–956.

Nelson, J., Buisine, S., Aoussat, A., & Gazo, C. (2014). Generating prospective scenarios of use in innovation projects. *Le Travail Humain,* numéro spécial "Ergonomie prospective", *77,* 21–38.

Nielsen, J. (1993). *Usability engineering.* Boston: Academic Press.

Osborn, A. F. (1963). *Applied imagination* (3rd ed.). New York: Scribner.

Pahl, G., Beitz, W., Feldhusen, J., & Grote, K.-H. (2007). *Engineering design—A systematic approach* (3rd ed.). Berlin: Springer.

Péladeau, P., Romac, B., Rozen, A., & Sevin, C. (2013). *L'innovation dans les entreprises en France*. Paris: Booz & Company Inc.

Picq, P. (2014). *Un paléoanthropologue dans l'entreprise*. Paris: Eyrolles.

Pruitt, J., & Adlin, T. (2006). *The persona lifecycle: Keeping people in mind throughout product design*. San Francisco: Morgan Kaufmann.

Ries, E. (2011). *The lean startup: How today's entrepreneurs use continuous innovation to create radically successful businesses*. New York: Crown Business.

Savransky, S. D. (2000). *Engineering of creativity: Introduction to TRIZ methodology of inventive problem solving*. Boca Raton, FL: CRC Press.

Schmitt, L., Buisine, S., Chaboissier, J., Aoussat, A., & Vernier, F. (2012). Dynamic tabletop interfaces for increasing creativity. *Computers in Human Behavior, 28*, 1892–1901.

Simon, H. A. (1973). The structure of ill-structured problems. *Artificial Intelligence, 4*, 181–201.

Tomiyama, T., Gu, P., Jin, Y., Lutters, D., Kind, C., & Kimura, F. (2009). Design methodologies: Industrial and educational applications. *CIRP Annals—Manufacturing Technology, 58*, 543–565.

Van Gundy, A. B. (2005). *101 activities for teaching creativity and problem solving*. San Francisco, CA: Wiley.

Vanderheiden, G. C. (1997). Design for people with functional limitations resulting from disability, aging and circumstance. In G. Salvendy (Ed.), *Handbook of human factors and ergonomics* (pp. 2010–2052). New York: Wiley.

Vanderheiden, G. C., & Tobias, J. (2000). Universal design of consumer products: Current industry practice and perceptions. In *Proceedings of the XIVth Triennal Congress of the International Ergonomics Association and 44th Annual Meeting of the Human Factors and Ergonomics Association* (pp. 19–22).

Vansteelandt, K., Mechelen, I. V., & Nezlek, J. (2005). The co-occurrence of emotions in daily life: A multilevel approach. *Journal of Research in Personality, 39*, 325–335.

Visser, W. (2009). La conception : de la résolution de problèmes à la construction de représentations. *Le Travail Humain, 72*, 61–78.

Von Hippel, E. (2005). *Democratizing innovation*. Cambridge, MA: MIT Press.

Yoshikawa, H. (1985). Design theory for CAD/CAM integration. *Annals of CIRP, 34,* 173–178.

Zelenski, J. M., & Larsen, R. (2000). The distribution of basic emotions in everyday life: A state and trait perspective from experience sampling data. *Journal of Research in Personality, 34,* 178–197.

8

Modelling the Creative Process in Design: A Socio-cognitive Approach

Julien Nelson

Introduction

Although design has long resisted attempts for a rigorous definition, it is often considered to be an archetypically creative activity. Design, today more than ever, is said to be at the forefront of the innovation economy (Le Masson, Weil, & Hatchuel, 2010). In this chapter, we focus more specifically on the creative process of designers, i.e. the sequence of thoughts and actions that leads designers to come up with a novel, adaptive production in response to a design brief (Lubart, 2001). Starting with Herbert Simon's call in the 1960s to inaugurate "a science of design" (Simon, 1969), we show that investigations of this creative process have considered creativity in design primarily as a psychological and social activity. We describe the main findings and limitations of each approach, and discuss future prospects for research.

J. Nelson (✉)
Paris Descartes University, Paris, France
e-mail: Julien.nelson@parisdescartes.fr

© The Author(s) 2018
T. Lubart (ed.), *The Creative Process*, Palgrave Studies in Creativity and Culture,
https://doi.org/10.1057/978-1-137-50563-7_8

Psychological Studies of Design: A Predominantly Cognitive Approach

Design as Symbolic Information Processing

The work on cognitive models of design is commonly said to have originated with Simon (1969) and his characterization of design as a complex, domain-general, problem-solving activity. More specifically, according to his and Allen Newell's seminal work on human problem solving (Newell & Simon, 1972), problem solving entails searching a problem space, which comprises an initial (i.e. current) state, a goal state, and a set of operators—or transformation functions—that can be used to move from the initial state to the goal state (Goel & Pirolli, 1989). Simon (1973) describes design problems as ill-structured, because they lack a number of key features, including definite criteria to test design solutions, a clear definition of the problem space and solution space, etc. In addition, design problems are also "wicked" (Rittel & Webber, 1973) because they present specific characteristics that make them difficult to solve: there is no definitive formulation of a given design problem, and there is never a clear stopping rule or "optimal solution". Instead, there are multiple acceptable solutions, and a design project (e.g. in architectural or product design) will usually end with the materialization of only one solution. As a result, every design project is unique, and it is difficult to derive knowledge that can be reused from one project to another.

In spite of this, many authors have sought to identify the cognitive processes underlying design activity, whether through case studies of design projects or experimental studies where participants—usually design students or professionals—carry out a standardized task in more or less tightly controlled conditions (Cross, 2001). Most often, such studies rely on protocol analysis methodology (Ericsson & Simon, 1984), i.e. where the participants' overt behaviours—e.g. sketching, gestures, verbalizations—are recorded and analysed. Early studies following this methodology include, for example, Akin (1978), Eastman (1970), and Lawson (1979).

One of the main findings of this line of research is the existence of a dynamic tension between formulation of the design problem, and elaboration of design solutions—a phenomenon known as *coevolution of problem and solution* (Dorst & Cross, 2001). Indeed, although a design project begins with a contractual document (the design brief), this document often describes the goals of the project only in vague terms (e.g. "design a flying car"). Part of the work involved in design consists of formulating additional constraints—i.e. conditions that need to be met by the design solution—in order to help the designer operate within a "constrained cognitive environment" (Bonnardel, 2000). Thus, Bonnardel makes a distinction between (a) prescribed constraints, which are explicitly formulated in the brief, (b) constructed constraints, which are inferred from these by the designer based on his/her expertise, and (c) deduced constraints, which are constraints added by the designer based on his/her analysis of the current state of the problem.

Another finding of this line of research relates to the generation of design solutions, more specifically the reuse of existing knowledge structures for this purpose. In an oft-quoted paper, Mayer (1989, p. 40) makes a distinction between routine and nonroutine problems: the first "*are familiar problems that, although not eliciting an automatic memorized answer, can be solved by applying a well-known procedure*", whereas the second "*are unfamiliar problems for which the problem solver does not have a well-known solution procedure and must generate a novel procedure*". In cases of routine design, it is possible that a solution to the design problem may be readily evoked from memory. However, in most design situations, a new solution must be generated from existing structures in memory (Visser, 1990). In this process, designers often construct an early version of the solution (Lawson, 1979), sometimes known as a "primary generator" (Darke, 1979), which can then serve as a basis both for future formulations of the problem and proposals of design solutions. For example, Bonnardel (2000) has argued persuasively that analogies play a crucial part in the generation of design solutions. Analogies consist of "*the transfer of knowledge from one situation to another by a process of mapping – finding a set of one-to-one correspondences (often incomplete) between aspects of one body of information and aspects of another*" (Gick & Holyoak, 1983, p. 2). In the case of design projects, analogies serve as a means for designers to transfer knowledge from past projects to new ones, by adapting past solutions and/or past problem formulations.

Design as a Reflective Conversation with the Situation

The reuse of available knowledge (i.e. "top-down" reasoning) does not explain all of design creativity. Schön (1983) proposed an alternate view, that of design as a "conversation with the situation". According to this view, design activity is characterized by a series of cycles termed "seeing-moving-seeing", in which the designer (1) extracts information from the current situation, (2) acts upon that situation, and (3) formulates an updated interpretation. This model has been supported by a substantial literature involving protocol studies of sketching in designers (Goldschmidt, 1991; Purcell & Gero, 1998; Schön & Wiggins, 1992; Suwa & Tversky, 1997). Sketches, from this point of view, serve as a means to externalize ideas generated throughout the design process, and to explore problem formulations and potential solutions by exploiting available degrees of freedom as shrewdly as possible.

Design as a Construction of Representations

As shown above, design cognition is characterized by an epistemological tension between two prevailing views: one derived from Simon's proposal to view design as a generic cognitive ability akin to "ill-structured problem-solving", and the other from Schön's proposal of a situated view, where interactions with the context serve as a primary driver for the creative design process. In an effort to reconcile these two views, Visser (2006a, 2006b, 2009) proposed two conceptual contributions. The first consists of describing design activity as *opportunistic*: design may proceed in a top-down fashion, following a hierarchical plan of action, but only for as long as this is the most cognitively efficient option. Designers may deviate at any time from a set plan of action, if it is not appropriate to a specific design situation. The second innovation consists of considering design as primarily focused on the construction of representations. These representations can be either external or internal. Internal, or mental representations are defined as circumstantial, goal-driven constructions, as opposed to knowledge structures, which are deemed to be more permanent (Bonnardel, 2006). More precisely, Visser (2006a, 2006b, 2009) identifies three key processes in design: the generation, transformation, and evaluation of representations. These occur until these representations *"are so concrete, detailed, and precise, that the resulting representation – the specifications of the artefact – specify explicitly and completely the implementation of the artefact"* (Visser, 2006b, p. 61).

Design Cognition or Design Psychology?

As the previous sections have shown, a considerable amount of literature has focused on describing the cognitive mechanisms of design at the individual level. The first remark that can be made concerns an overreliance on Simon's paradigm of bounded rationality, and its relevance to describing the creative aspects of design (Hatchuel, 2001). Indeed, every design problem has a theoretically infinite number of solutions, and designing involves starting with an initial concept—usually contained in the design brief—and proposing interesting expansions of that concept. This line of thinking has led to a renewed interest in obstacles to this expandable rationality—in particular design fixation, defined as *"a blind, and sometimes counterproductive, adherence to a limited set of ideas in the design process"* (Jansson & Smith, 1991)—and in the organizational means to counter them. For example, Agogué et al. (2014) have studied the effects of providing different types of examples of design solutions on creative output, whereas Hooge and David (2014), using a case study methodology, have studied the effect of the manner of formulation of a theme for a creative workshop on creative performance and participant satisfaction.

In addition to these issues concerning the type of paradigm that might best characterize design cognition, it should be noted that these studies focus mainly on design as an individual task. However, real-world design practice implies often collaboration between individuals working in groups. Hence, it is necessary to consider the creative process in design, not just from a cognitive perspective, but from a social one as well. This will be addressed in the next section.

Design as a Social Undertaking

An Increased Demand for Collaboration in Design

From the literature outlined above, it should come as no surprise that creative design is a risky enterprise. The creative output of a designer, in response to a design brief, may not be novel, or it may be inappropriate to the client's (or consumer's) expectations or needs. To mitigate this risk, the 1980s and 1990s saw the emergence of new practices in design management. Acknowledging the fact that successful innovation required transfer of knowledge between experts of different fields, several authors (see in particular Takeuchi & Nonaka, 1986; Wheelwright & Clark, 1992) advocated the rise of project-based management. In such a work organization, the company's innovative ventures are placed under the responsibility of cross-functional teams, where experts from different fields must simultaneously learn from one another and act towards a common goal. Naturally, this emphasis on communication and coaction in teams is a major challenge in such teams, and this led in turn to an increased interest in the cognitive aspects of collaborative design—largely in terms of construction of external and internal representations (see "Design as a Construction of Representations").

The first source of complexity of collaboration in design projects is the variety of forms of collaboration that it entails. A fundamental distinction is often made between (a) *co-design*, where group members—collectively referred to as designers—are jointly involved in completing the same design task, each contributing his/her own expertise; and (b) *distributed design*, where they contribute to the same overall project, each completing separate allocated tasks (Darses & Falzon, 1996). Hence, participants are engaged in a process of twofold synchronization: *operative* synchronization serves to allocate interdependent tasks between group members, whereas *cognitive* synchronization refers to establishing a "common ground" between them. This comprises a network of shared information regarding mutual knowledge, beliefs and assumptions (Clark & Brennan, 1991). This common frame of reference concerns *"the current state of the problem, solution, plans, design rules and more general design knowledge"* (Détienne, 2006, p. 11), and is regularly updated throughout collaboration.

The second source of complexity relates to integrating the multiple viewpoints of designers so that the design process can converge towards a single solution. Several authors in recent years have taken an interest in the mechanisms of argumentation and negotiation that make this convergence possible (Darses, 2006; Détienne et al., 2016). Such studies rely often on an ethnographic approach, examining primarily verbal interactions between participants, and how these interactions make it possible, in particular, to (1) generate propositions regarding the object to be designed, (2) formulate arguments in favour of—or against—these propositions, and (3) evaluate these propositions. Creative design is seen here as a social process (Bucciarelli, 1988).

From a Cognitive to an Object-Centred Approach of Collaboration in Design

Following this approach, a design project is an organizational framework that allows professionals of different trades to contribute to the elaboration of a single creative product. In doing so, they produce artefacts of specific *types* with which they interact in specific *ways*. These interactions of individual designers with the artefacts produced in their work had already been explored within a situated approach (see "Design as a Reflective Conversation with the Situation"). However, later work, also based on an ethnographic approach of design situations, examined how these objects structure creative collaboration in groups of designers. Referring to Actor Network Theory (e.g. Latour & Woolgar, 1979) and the concept of *boundary object* (Star & Griesemer, 1989), Vinck and Jeantet (1995) introduced the concept of "Intermediary Object of Design" (IOD). These objects are defined as *"representations of a final, absent object. They are supposed to be objects that can be communicated and exchanged between design partners"* (Vinck, Jeantet, & Laureillard, 1996). This concept has enjoyed considerable success in ethnographic studies of design and engineering practices (Boujut & Blanco, 2003; Vinck, 2003). It has also led to a renewal in these studies as it places much more emphasis on the objects produced throughout the design process, how designers interact with them, and how these objects and interactions structure the creative design process.

Elsen (2011) lists the types of IODs that are most commonly used in industrial design: in addition to classical tools for communication (e.g. telephone, emails) and tools used for creative and artistic research (e.g. painting, brainstorming, etc.), she notes that several objects are used to structure the creative process per se, corresponding to the gradual progression from the initial idea to the finished product: sketches and annotations (on plain paper or in digital form), files produced from Computer-Aided Design (CAD) software, mock-ups, and functional prototypes. Her analysis, which focuses on a comparison between pen-and-paper and digital tools, points out that these two families of tools do not support the creative design process in the same way. Pen-and-paper drawings serve as a means to externalize creative ideas, but also to develop them in flexible ways: just as designers may interact with their own sketches (e.g. adding or correcting information), they may use a sketch as a device for communicating with other people and negotiation of design choices, but also as archives of past design alternatives. A sizeable literature has studied the creative functions of sketches and interactions with them (e.g. adding or correcting lines, varying thickness, hatching, colouring, annotations, etc.). These various traces serve to maintain flexibility and progressively validate the most relevant creative choices. However, each designer will exhibit specific practices in both the production and interpretation of sketches.

In contrast, CAD files make use of the potential of computational tools to automatize, in part or completely, some aspects of design and manufacturing. In particular, two main aspects can benefit from this potential: (1) producing detailed design solutions and evaluating technical aspects of these solutions—e.g. thermal or structural properties—following formal rules and (2) producing realistic visual renderings of design solutions for the purposes of communication. This second point is particularly important, as CAD software makes it possible to assess design solutions without the need to produce physical objects (e.g. prototypes). However, the use of this type of software forces designers to produce immediately detailed design solutions, leading to fixation effects. In more recent years, several authors have sought to explore the potential of combining the advantages of these two modes of creative expression—pen-and-paper vs. CAD based design (Ben Rajeb & Leclercq, 2015; Safin, 2012). The ambition of these studies is to propose a fully digitized design process, one where the judicious combination of analog and digital media—i.e. allowing the flexible generation and evaluation of early as well as detailed design solutions, by individuals as well as groups.

From Creative to Managerial Processes ... and Back

A Need to Mitigate the Risks of Creative Design Through Prescriptive Models

It should be clear at this point that creative design is a risky enterprise. Indeed, the design process can produce output that is not novel, not suitable to task constraints, or both. The consequences of such failures of creativity can be disastrous, as companies rely on innovation to thrive and survive (Le Masson et al., 2010). Hence, the concept of "design project" echoes a simple requirement for companies: to ensure that a team of professionals produce a high-quality output while enforcing three main success criteria: time, costs, and quality (Atkinson, 1999). Furthermore, because design can be thought of as a coevolution between problem and solution, the end product of the design project cannot be thought of as the answer to just one problem, but to the entire set of problems that emerged during the project. A design project, then, implies a gradual process of convergence towards a single solution (Midler, 1995).

This framework implies typically following a prescriptive model to structure design work. In reference to the tradition of activity-centred ergonomics (Daniellou, 2005), this terminology refers to a model of (design) work *as it should be done*, usually according to the work organization, with the understanding that this formal model often differs from how work is actually conducted. In a review of 23 prescriptive models of the engineering design process proposed between 1967 and 2006, Howard, Culley, and Dekoninck (2008) noted a consensus between most models on an overall linear structure comprising four main stages: analysis of task, conceptual design, embodiment design, and detailed design. This terminology is notably similar to that proposed by Pahl and Beitz (1977), one of the most influential models of engineering design process. According to these authors, task analysis (termed clarification of the task in the model) is concerned with collecting information about the requirements and constraints that apply to the design solution. Conceptual design "involves the establishment of function structures, the search for suitable solution principles, and their combination into concept variants" (p. 40). Embodiment design aims, starting from a design concept, to produce a layout, i.e. to develop "a technical product, or system, in accordance with technical and economic considerations" (p. 42). Finally, detailed design is "the phase of the design process in which the arrangement, form, dimensions, and surface properties of all the individual parts are finally laid down, their materials specified, the technical and economic feasibility rechecked, and all the drawings and other production documents produced" (p. 42). These four stages do indeed correspond to an increasing level of knowledge relative to the object being designed, and to increasingly specific design questions. However, Howard et al. (2008) include also two further stages in their analysis: the "establishing a need phase", before the analysis of the design task, aims to identify an opportunity to launch a design project. The implementation phase, after the detailed design phase, consists of releasing the finished product (e.g. in the market, in the workplace). This expansion reveals an important evolution: rather than to ensure the success of a single design project, design management today aims to channel the available creative potential towards (a) the creation of multiple products and (b) their successful introduction to a market of users (Le Masson et al., 2010).

From a Cognitive to a Socio-cognitive Approach

As noted by Botella, Nelson, & Zenasni (2016), the creative process can be described at two levels: Macro level processes describe the successive stages of the creative work, usually in terms of task goals or actions undertaken to achieve them. Micro level processes describe the mechanisms involved in these stages.

Early research on creativity, particularly Guilford (1959) emphasized the importance of divergent thinking or the ability to "think in different directions" in response to a problem. Convergent thinking, on the other hand, is defined as the ability to produce one single answer. As noted in "Design Cognition or Design Psychology?" , Simon's paradigm of bounded rationality reminds us that design does not aim to identify the single best solution to a design problem, but instead to identify an acceptable ("satisficing") solution, i.e. one that best responds to the constraints posed by the design brief and defined by the designers themselves. Several models of the design process have argued that design involved both a succession of creative divergence and convergence, and global convergence towards an end product. However, these cycles of divergence and convergence occur at a macro level, and should not be confused with cycles of divergent and convergent *thinking*. A more apt term, proposed by reference, would be generation and evaluation of local solutions. The latter topic has gathered significant interest in recent years (see e.g. Bonnardel & Sumner, 1996; Wojtczuk & Bonnardel, 2011). However, this phenomenon is usually studied in the context of individual designers, or groups of designers interacting in the context of a design meeting.

Recent studies, in the field of Computer-Supported Cooperative Work, have sought to account for the social aspects of design projects following a situated cognition perspective (Barcellini, Détienne, & Burkhardt, 2014; Barcellini, Détienne, Burkhardt, & Sack, 2008). The authors chose to study a large Open-Source Software Design (OSSD), Python. This project involves large number of participants interacting in online spaces. Their interactions take place in dedicated discussion spaces, making it possible to access past discussions for detailed analysis. The social component is present at two levels: (a) the existence of a core-periphery hierarchy that is native to this particular online community and (b) the actual activities—or *roles*—performed by project participants which define their role within a specific collaboration space. It is therefore possible to study the creative process by analysing the structure and content of the online discussions in which participants perform the activities described above: problem analysis and reformulation, solution generation and evaluation, before implementation of design solutions in the software. This kind of analysis makes it possible to identify participant profiles, i.e. distributions of roles across multiple discussion spaces, thereby reconstructing their effective contribution to the design process. This kind of study, aiming to describe the co-elaboration of a design object in terms of situated cognition as well as social structuration, has already been used in several different contexts, including most recently the design of software for use in agricultural practice (Barcellini, Prost, & Cerf, 2015) the co-elaboration of knowledge on Wikipedia (Détienne et al., 2016).

Discussion: Design or Creative Design?

In this chapter, we sought to examine the creative processes underlying design following two distinct points of view:

• From a cognitive point of view, Herbert Simon's initial proposal for establishing a "science of design" has proven very influential, but his qualification of design in terms of ordinary cognitive processes has been criticized for not being able to convey the more creative aspects of design activity. Creativity, at the individual level, emerges when the designer (a) utilizes or combines existing knowledge structures in novel and suitable ways, (b) reformulates the initial design problem, represented in the brief, so that creative solutions may emerge, and (c) produces novel and adapted solutions to these problems. Current discussions seem to be focused on alleviating the cognitive and/or organizational obstacles to creativity (design fixation). Numerous techniques of creative design exist to achieve this, but surprisingly few have been the subject of experimental or ethnographic studies.

- From a social point of view, the acknowledgement that most design projects are intrinsically collaborative has led to an increased interest in the study of these collaborations. The focus of these analyses is on interactions between individual designers and their environment, be it social (i.e. with other people, most often members of the design team) or material (i.e. with external objects that serve to mediate a conversation, both reflective and argumentative, concerning the object that is being designed). Studies following a socio-cognitive approach (see "From a Cognitive to a Socio-cognitive Approach") are typically ethnographic in nature and focus on how social interactions within a group support the co-elaboration of a design object. However, they address only the interactional processes, not the quality of the ideas expressed in individual contributions (Barcellini et al., 2014).

From these two points, it is easy to see that studies of design activity often account only partly for the creative processes in design, notably because the creativity of the design outcome is not assessed. Although the creativity literature has proposed methods to assess the intrinsic quality of creative ideas—the most notable of which being the Consensual Assessment Technique (Amabile, 1982)—the ultimate judgement of a creative production can only be made in reference to a social context. This context makes it possible, on the one hand, to differentiate eminent instances of creative production from everyday creative achievements (Kaufman & Beghetto, 2009). But, especially in the context of design, this is not enough. The very production and implementation of a creative product, its ability to reach its expected target, is dependent on the ability of an original idea (and its producer) to convince people and steer the context in its favour (Akrich, Callon, & Latour, 2002; Sternberg & Lubart, 1995). This element is part and parcel of the creative process, yet the underlying social interactions remain largely unknown (Bucciarelli, 1994): with whom must a designer interact to ensure that an idea gets "patched through" to the next stage of design? How are these interlocutors convinced? Clearly, there are complex social processes at work, which may occur outside of the design team, but may in turn impact the creative output of the designers themselves. Furthermore, it is likely that these interactions may not just be evaluative, but also generative in nature. Innovations in the design process have already called for design to move out of the designer's office and for non-designers to adopt "design thinking" (Brown, 2008). Evolutions in investigation methods will make it possible to track the creative process in these new forms of design-oriented organizations.

On the contrary, contemporary management theory posits that the success of a project relies on the engagement of multiple stakeholders (Freeman, 1984). The success of a design project, in terms of innovation, depends on the ability of an idea to gather strategic allies (Akrich et al., 2002). For example, in the field of product design, the success of a project will, of course, be determined in no small part by how the end users of the product evaluate it, and subsequently behave—i.e. adopt or reject it. End users can also be mobilized to propose new innovations (e.g. von Hippel, 2005). Hence, an existing design output can serve as a starting point for further creative achievements after a project has finished. Less emphasis is placed, however, on the fact that the assessment of creative products is also involved in earlier stages of the design process. For example, before a company invests resources in a design project, designers must first convince stakeholders within the company that their idea constitutes an interesting source of potential value to that company. In this sense, design is a creative activity not just during the creation of a new artefact, but also in the stages preceding and following that creation.

References

Agogué, M., Kazakçi, A., Hatchuel, A., Le Masson, P., Weil, B., Poirel, N., & Cassotti, M. (2014). The impact of type of examples on originality: Explaining fixation and stimulation effects. *Journal of Creative Behavior*, *48*(1), 1–12. http://doi.org/10.1002/jocb.37.

Akin, O. (1978). How do architects design? In J. C. Latombe (Ed.), *Artificial intelligence and pattern recognition in computer-aided design* (pp. 65–104). New York, NY: North Holland.

Akrich, M., Callon, M., & Latour, B. (2002). The key to success in innovation, part 1: The art of intéressement. *International Journal of Innovation Management, 6*(2), 187–206. http://doi.org/10.1142/S1363919602000550.

Amabile, T. M. (1982). Social psychology of creativity: A consensual assessment technique. *Personality Processes and Individual Differences, 43*(5), 997–1013.

Atkinson, R. (1999). Project management: Cost time and quality two best guesses and a phenomenon, it's time to accept other success criteria. *International Journal of Project Management, 17*(6), 337–342. https://doi.org/10.1016/S0263-7863(98)00069-6.

Barcellini, F., Détienne, F., & Burkhardt, J. M. (2014). A situated approach of roles and participation in open source software communities. *Human–Computer Interaction, 29*(3), 205–255. https://doi.org/10.1080/07370024.2013.812409.

Barcellini, F., Prost, L., & Cerf, M. (2015). Designers' and users' roles in participatory design: What is actually co-designed by participants? *Applied Ergonomics, 50,* 31–40. http://doi.org/10.1016/j.apergo.2015.02.005.

Barcellini, F., Détienne, F., Burkhardt, J. M., & Sack, W. (2008). A socio-cognitive analysis of online design discussions in an open source software community. *Interacting with Computers, 20*(1), 141–165. https://doi.org/10.1016/j.intcom.2007.10.004.

Ben Rajeb, S., & Leclercq, P. (2015). Instruments for collective design in a professional context: Digital format or new processes? In *Eighth International Conference on Advances in Computer-Human Interactions.*

Bonnardel, N. (2000). Towards understanding and supporting creativity in design: Analogies in a constrained cognitive environment. *Knowledge-Based Systems, 13*(7–8), 505–513. http://doi.org/10.1016/S0950-7051(00)00067-8.

Bonnardel, N. (2006). *Créativité et conception: Approches cogntiives et ergonomiques.* Marseille: Solal.

Bonnardel, N., & Sumner, T. (1996). Supporting evaluation in design. *Acta Psychologica, 91*(3), 221–244.

Boujut, J.-F., & Blanco, E. (2003). Intermediary objects as a means to foster co-operation in engineering design. *Computer Supported Cooperative Work (CSCW), 12*(2), 205–219. https://doi.org/10.1023/A:1023980212097.

Botella, M., Nelson, J., & Zenasni, F. (2016). Les macro- et microprocessus créatifs. In I. Capron Puozzo (Ed.), *La créativité en éducation et en formation: Perspectives théoriques et pratiques* (pp. 33–43). Louvain-la-Neuve: De Boeck.

Brown, T. (2008). Design thinking. *Harvard Business Review, 86*(6). http://doi.org/10.1145/2535915.

Bucciarelli, L. L. (1994). *Designing engineers. Inside technology* (Vol. 25). http://doi.org/10.1080/03043799508928289.

Bucciarelli, L. L. (1988). An ethnographic perspective on engineering design. *Design Studies, 9*(3), 159–168.

Clark, H. H., & Brennan, S. (1991). Grounding in communication. In L. B. Resnick, J. M. Levine, & S. D. Teasley (Eds.), *Perspectives on socially-shared cognition* (pp. 127–149). Washington, DC: American Psychological Association. http://doi.org/10.1037/10096-006.

Cross, N. (2001). Design cognition: Results from protocol and other empirical studies of design activity. *Design Knowing and Learning: Cognition in Design Education*, 79–104. http://doi.org/10.1016/B978-008043868-9/50005-X.

Daniellou, F. (2005). The French-speaking ergonomists' approach to work activity: Cross-influences of field intervention and conceptual models. *Theoretical Issues in Ergonomics Science, 6*(5), 409–427. http://doi.org/10.1080/14639220500078252.

Darke, J. (1979). The primary generator and the design process. *Design Studies, 1*(1), 36–44. https://doi.org/10.1016/0142-694X(79)90027-9.

Darses, F. (2006). Analyse du processus d'argumentation dans une situation de reconception collective d'outillages. *Le Travail Humain, 69*(4), 317–347. https://doi.org/10.3917/th.694.0317.

Darses, F., & Falzon, P. (1996). La conception collective: une approche de l'ergonomie cognitive. In G. de Terssac & E. Friedberg (Eds.), *Coopération et conception* (pp. 123–135). Toulouse: Octarès.

Détienne, F. (2006). Collaborative design: Managing task interdependencies and multiple perspectives. *Interacting with Computers, 18*(1 spec. iss.), 1–20. http://doi.org/10.1016/j.intcom.2005.05.001.

Détienne, F., Baker, M., Fréard, D., Barcellini, F., Denis, A., & Quignard, M. (2016). The descent of Pluto: Interactive dynamics, specialisation and reciprocity of roles in a Wikipedia debate. *International Journal of Human Computer Studies, 86*, 11–31. https://doi.org/10.1016/j.ijhcs.2015.09.002.

Dorst, K., & Cross, N. (2001). Creativity in the design process: Co-evolution of problem–solution. *Design Studies, 22*(5), 425–437. https://doi.org/10.1016/S0142-694X(01)00009-6.

Eastman, C. M. (1970). On the analysis of intuitive design processes. In G. T. Moore (Ed.), *Emerging methods in environmental design and planning* (pp. 21–37). Cambridge, MA: MIT Press.

Elsen, C. (2011). *La médiation par les objets en design industriel: Perspectives pour l'ingénierie de conception.* Ph.D. thesis, University of Liège, Liège.

Ericsson, K. A., & Simon, H. A. (1984). *Protocol analysis: Verbal reports as data.* Cambridge, MA: Bradford Books.

Freeman, R. E. (1984). *Strategic management: A stakeholder approach.* Boston, MA: Pitman.

Gick, M. L., & Holyoak, K. J. (1983). Schema induction and analogical transfer. *Cognitive Psychology, 15*(1), 1–38. https://doi.org/10.1016/0010-0285(83)90002-6.

Goel, V., & Pirolli, P. (1989). Motivating the notion of generic design with information processing theory: The design problem space. *AI Magazine, 10*(1), 18–36.

Goldschmidt, G. (1991). The dialectics of sketching. *Creativity Research Journal, 4*(2), 123–143. https://doi.org/10.1080/10400419109534381.

Guilford, J. P. (1959). Three faces of intellect. *American Psychologist, 14*(8), 469–479. https://doi.org/10.1037/h0046827.

Hatchuel, A. (2001). Towards design theory and expandable rationality: The unfinished program of Herbert Simon. *Journal of Management and Governance, 5*(3–4), 260–273. https://doi.org/10.1023/A:1014044305704.

Hooge, S., & David, A. (2014). What makes an efficient theme for a creativity session? XXIth International Development Management Conference (IPDMC), June 2014, Limerick, Ireland.

Howard, T. J., Culley, S. J., & Dekoninck, E. (2008). Describing the creative design process by the integration of engineering design and cognitive psychology literature. *Design Studies, 29*(2), 160–180. https://doi.org/10.1016/j.destud.2008.01.001.

Jansson, D. G., & Smith, S. M. (1991). Design fixation. *Design Studies, 12*(1), 3–11. https://doi.org/10.1016/0142-694X(91)90003-F.

Kaufman, J. C., & Beghetto, R. A. (2009). Beyond big and little: The four c model of creativity. *Review of General Psychology, 13*(1), 1–12. https://doi.org/10.1037/a0013688.

Latour, B., & Woolgar, S. (1979). *Laboratory life: The social construction of scientific fact*. Beverly Hills, CA: Sage.

Lawson, B. (1979). Cognitive strategies in architectural design. *Ergonomics, 22*(1), 59–68. https://doi.org/10.1080/00140137908924589.

Le Masson, P., Weil, B., & Hatchuel, A. (2010). *Strategic management of innovation and design*. Cambridge, UK: Cambridge University Press.

Lubart, T. I. (2001). Models of the creative process: Past, present and future. *Creativity Research Journal*. http://doi.org/10.1207/S15326934CRJ1334_07.

Mayer, R. E. (1989). Human nonadversary problem-solving. In K. J. Gilhooly (Ed.), *Human and machine problem solving* (pp. 39–56). New York, NY: Plenum.

Midler, C. (1995). "Projectification" of the firm: The Renault case. *Scandinavian Journal of Management, 11*(4), 363–375. https://doi.org/10.1016/0956-5221(95)00035-T.

Newell, A., & Simon, H. A. (1972). *Human problem solving*. Englewood Cliffs, NJ: Prentice Hall.

Pahl, G., & Beitz, W. (1977). *Engineering design: A systematic approach*. Berlin: Springer Verlag.

Purcell, A. T., & Gero, J. S. (1998). Drawings and the design process. *Design Studies, 19*(4), 389–430. https://doi.org/10.1016/S0142-694X(98)00015-5.

Rittel, H. W. J., & Webber, M. M. (1973). Dilemmas in a general theory of planning. *Policy Sciences, 4*(2), 155–169. https://doi.org/10.1007/BF01405730.

Safin, S. (2012). Use of graphical modality in a collaborative design distant setting. In J. Dugdale, C. Masclet, M. A. Grasso, J. F. Boujut, & P. Hassanaly (Eds.), *Proceedings of the 10th international conference on the design of cooperative systems* (Vol. 2, pp. 245–261). London: Springer.

Schön, D. A. (1983). *The reflective practitioner: How professionals think in action*. New York, NY: Basic Books.

Schön, D. A., & Wiggins, G. (1992). Kinds of seeing in designing. *Creativity and Innovation Management, 1*(2), 68–74. https://doi.org/10.1111/j.1467-8691.1992.tb00031.x.

Simon, H. A. (1969). *The sciences of the artificial* (1st ed.). Cambridge, MA: MIT Press.

Simon, H. A. (1973). The structure of ill structured problems. *Artificial Intelligence, 4*(3–4), 181–201. https://doi.org/10.1016/0004-3702(73)90011-8.

Star, S. L., & Griesemer, J. R. (1989). Institutional ecology, 'translations' and boundary objects: Amateurs and professionals in Berkeley's Museum of Vertebrate Zoology, 1907–39. *Social Studies of Science, 19*(3), 387–420. https://doi.org/10.1177/030631289019003001.

Sternberg, R. J., & Lubart, T. I. (1995). *Defying the crowd: Cultivating creativity in a culture of conformity*. New York, NY: Free Press.

Suwa, M., & Tversky, B. (1997). What do architects and students perceive in their design sketches? A protocol analysis. *Design Studies, 18*(4), 385–403. http://doi.org/10.1016/S0142-694X(97)00008-2.

Takeuchi, H., & Nonaka, I. (1986). The new new product development game. *Journal of Product Innovation Management, 3*(3), 205–206. https://doi.org/10.1016/0737-6782(86)90053-6.

Vinck, D. (Ed.). (2003). *Everyday engineering: An ethnography of design and innovation*. Cambridge, MA: MIT Press.

Vinck, D., & Jeantet, A. (1995). Mediating and commissioning objects in the sociotechnical process of product design: A conceptual approach.

In D. MacLean, P. Saviotti, & D. Vinck (Eds.), *Management and new technology: Design networks and strategies* (Vol. 2, pp. 111–129). Brussels: EC Directorate General Science R&D.

Vinck, D., Jeantet, A., & Laureillard, P. (1996). Objects and other intermediaries in the sociotechnical process of product design : An exploratory approach. In EC Directorate (Ed.), *The role of design in the shaping of technology* (Vol. 5, pp. 297–320). Brussels.

Visser, W. (1990). Evocation and elaboration of solutions: Different types of problem-solving actions. An empirical study on the design of an aerospace artifact. In *Cognitiva 90. At the crossroads of artificial intelligence, cognitive science, and neuroscience. Proceedings of the Third Cognitiva Symposium* (pp. 689–696). Amsterdam: Elsevier.

Visser, W. (2006a). *Cognitive artifacts of designing*. Mahwah, NJ: Lawrence Erlbaum Associates.

Visser, W. (2006b). Designing as construction of representations: A dynamic viewpoint in cognitive design research. *Human–Computer Interaction, 21*(1), 103–152.

Visser, W. (2009). Design: One, but in different forms. *Design Studies, 30*(3), 187–223. https://doi.org/10.1016/j.destud.2008.11.004.

von Hippel, E. (2005). *Democratizing innovation*. Cambridge, MA: MIT Press.

Wheelwright, S. C., & Clark, K. B. (1992). *Revolutionizing product development: Quantum leaps in speed, efficiency, and quality*. New York, NY. http://doi.org/citeulike-article-id:4107539.

Wojtczuk, A., & Bonnardel, N. (2011). Designing and assessing everyday objects: Impact of externalisation tools and judges' backgrounds. *Interacting with Computers, 23*(4), 337–345. https://doi.org/10.1016/j.intcom.2011.05.004.

9

The Creative Process in Design

Nathalie Bonnardel, Alicja Wojtczuk, Pierre-Yves Gilles
and Sylvain Mazon

Introduction

Companies and designers face the challenge of coming up with products that are both new and adapted to future users. The process of
design thinking therefore requires creativity. However, the task of
imagining and conceiving new products is particularly complex for
designers (Bonnardel, 2012; Bonnardel, Forens, & Lefevre, 2016). In
order to understand the process of design thinking more clearly, we
begin by describing the general characteristics and current models of
both creativity and design activities, at the macro and microprocess

N. Bonnardel (✉) · A. Wojtczuk · P.-Y. Gilles · S. Mazon
PSYCLE, Aix Marseille University, Aix-en-Provence, France
e-mail: nathalie.bonnardel@univ-amu.fr

P.-Y Gilles
e-mail: pierre-yves.gilles@univ-amu.fr

S. Mazon
e-mail: sylvain.mazon@orange.fr

© The Author(s) 2018
T. Lubart (ed.), *The Creative Process*, Palgrave Studies in Creativity and Culture,
https://doi.org/10.1057/978-1-137-50563-7_9

levels ("Characteristics and Models of the Process of Creative Design Thinking"). We then describe two studies we conducted: the first with professional designers to determine how they perceive their own process of creative design thinking ("General Characteristics of Design Activities"), and the second with design students to determine how teaching methods based on either divergent or convergent thinking influence certain features of their design thinking process ("Macroprocess of Creative Design Thinking"). Finally, in the light of these studies, we identify characteristics of the creative design process that could be enhanced by particular teaching and/or computational modalities ("Microprocesses Involved in Creative Design Thinking").

Characteristics and Models of the Process of Creative Design Thinking

General Characteristics of Design Activities

From a cognitive point of view, design activities can be regarded as problem-solving activities. More specifically, we adhere to Treffinger (1995)'s definition of a *problem* as any "important, open-ended, and ambiguous situation for which one wants and needs new options and a plan for carrying a solution successfully" (p. 304). Design problems are regarded as *ill structured* or *ill defined* (Eastman, 1969; Simon, 1995), insofar as designers' mental representations are initially incomplete and imprecise. These representations therefore have to be completed through the coevolution of problem and solution spaces (Dorst & Cross, 2001). Moreover, when designers search for a design solution, they engage frequently in an *opportunistic process*, where each new decision is motivated by the one before (Hayes-Roth, 1979; Visser, 1994). As a consequence, the decision process results from both top-down and bottom-up processes. It can be multidirectional and occur at different levels of abstraction.

Design problems are also regarded as *open-ended* or *wicked* problems, insofar as they have a large number of possible solutions (Rittel & Webber, 1984). Furthermore, as pointed out previously, in creative (or non routine) design contexts, design solutions must be both *new* and *adapted* to the characteristics of the situation or context, including future users and usages (Bonnardel, 2006, 2012). This twofold requirement corresponds directly to definitions of creativity, which can be viewed as the ability to produce work that is both novel and appropriate (Sternberg & Lubart, 1999). In accordance with this definition, in design situations, we argue that a creative approach (Bonnardel, 2000, 2006) requires both the enlargement of the search area for creative ideas and a focus on the project's constraints. Designers must come up with original products that are quite distinct from existing ones, but which suit and do not destabilize users. Creativity is therefore dependent on both the individual who creates the new products and the environment and society in which these products are created (Csíkszentmihályi, 1996; Lubart, Mouchiroud, Tordjman, & Zenasni, 2003).

Macroprocess of Creative Design Thinking

According to Botella, Nelson, and Zenasni (2016), two approaches can be adopted to describe the creative process. One allows for the description of the different stages of the *macroprocess*, whereas the other allows for the identification of the mechanisms (or *microprocesses*) involved in the creation of ideas. These two types of descriptions suggest that creativity can be analyzed as either one general process or several specific processes, depending on the level of analysis and the characteristics of the individual or context.

At the macro level of the design process, early models described it as being based on *a sequence of stages*. Asimov (1962), for instance, identified three such stages: analysis, synthesis, and evaluation (see Fig. 9.1). *Analysis* corresponds to gathering the relevant information (or preparing the problem) and framing (or (re)formulating) the problem; *synthesis* is associated with the search for an appropriate solution; and *evaluation* can be described as the validation of the proposed solution.

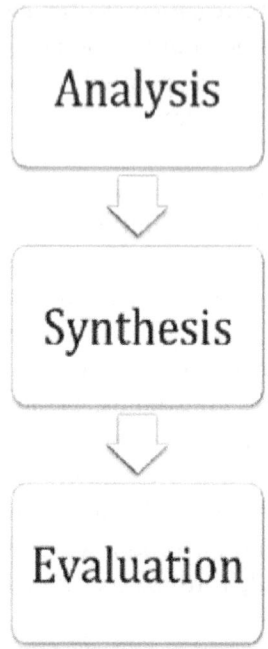

Fig. 9.1 Design process by Asimov (1962)

If the evaluation stage yields unsatisfactory results, the whole process is repeated. This model was reinterpreted by McNeil, Gero, and Warren (1998), who showed that there is more than one possible sequence for these stages: after evaluation, for instance, the designer can proceed to either the analysis or the synthesis stage.

Bonnardel (2009) compared these models of the design process with models of the creative process, such as those developed by Wallas (1926) and Amabile (1996) (see Figs. 9.2 and 9.3). For example, Wallas (1926) pointed out the importance of an incubation phase, during which numerous associations are made without any conscious effort on the designer's part. These associations may be followed by some kind of illumination, when an interesting idea is consciously considered. Subsequent models of the creative process added several layers of complexity to this initial proposal. Amabile (1996), for example, included

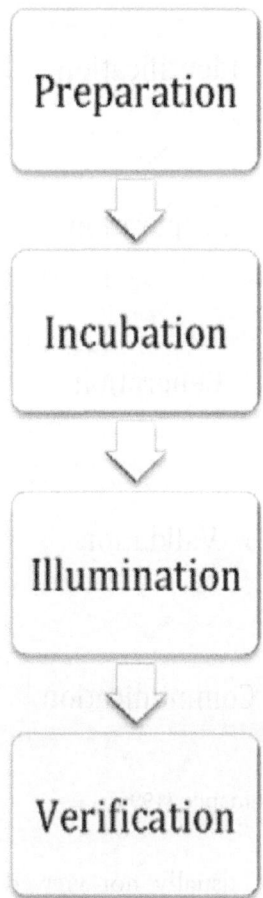

Fig. 9.2   Creative process by Wallas (1926)

other stages in the creative process, such as identifying the problem and communicating the final output to others. Mumford, Mobley, Reiter-Palmon, Uhlman, Doares (1991) also described a more elaborate sequence of stages (problem construction, information encoding, category search, specification of best-fitting categories, idea evaluation, implementation, monitoring), with many feedback loops between them.

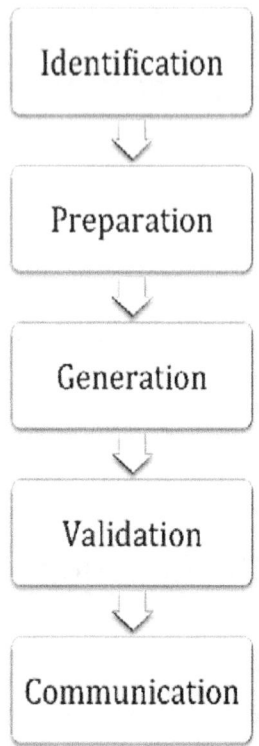

Fig. 9.3 Creative process by Amabile (1996)

Designers' activities are usually not viewed as a linear succession of stages, but rather as an iterative process (as described in "General Characteristics of Design Activities"), in which the mental representation of the (initially incomplete) problem gradually comes into sharper focus as the problem-solving process moves forward (Simon, 1973, 1995). Zeisel (1981)'s metaphor of a spiral is a good illustration of the dynamics of the design problem-solving process and its conceptual jumps between iterative cycles. Schön (1983) talked about a *reflective conversation* between the designer and external representations of the object to be designed. During this process, designers make unexpected discoveries, which can be positive (results perfectly meeting requirements), negative (emergence of problems interfering with the goals being pursued), or innovative (perception of new directions for creative search). Other models of design activity include components related to situated cognition. This is, especially, the case of the function–behavior-structure (FBS) model (Gero, 1998), the analogy and constraint management (A-CM) model (Bonnardel, 2000), and the description of design put forward by Tan and Melles (2010), in line with the opportunistic nature of design activities (see "General Characteristics of Design Activities").

These models were recently complemented by analyses and comparisons between different creative domains (Glăveanu et al., 2013), with reference to John Dewey's work (1934). Creativity is based on the interaction between a person and the environment, and is intrinsically related to human activity with and within the world (for more information, see Glăveanu, 2012). According to Glăveanu et al. (2013), action is a continuous cycle of *doing* (actions directed toward the environment) and *undergoing* (taking in the reaction of the environment). Through these interconnected processes, action can be taken forward so that it becomes a *full* experience.

Microprocesses Involved in Creative Design Thinking

In accordance with the dual criteria of novelty and adaptation to the context, the A-CM model (Bonnardel, 2000, 2006) highlights the roles of two main cognitive processes, which may have contrasting effects and contribute to both divergent and convergent thinking.

1. *Analogical thinking* and, more generally, *associations* can, in certain circumstances (see, for instance, Bonnardel & Marmèche, 2004, 2005), lead designers to open up or extend their search space to new ideas, in line with divergent thinking. Although they can engage in other forms of creative thought (see, for instance, Mumford, 2003), we argue that it is important for designers to establish connections between the design domain (e.g., a mechanical device) and other domain(s) that can provide them with inspiration (e.g., a biological system). Analogical thought can allow designers to transfer the

features of one or more sources of inspiration to the design solution being constructed. Final design solutions can therefore result from the combination of several concepts or ideas. Associations can also be useful when designers wish to propose solutions whose features contrast with the source of inspiration and thus break with preexisting objects, products or entities (Bonnardel, Didierjean, & Marmèche, 2003). However, there is also a risk of *design fixation*. If designers are too focused on sources of inspiration, they may have difficulty opening up their idea search space (Chrysikou & Weisberg, 2005; Jansson & Smith, 1991). Thus, conditions have been identified that allow designers to benefit from analogical thinking and associations during the process of design thinking (see, for instance, Bonnardel & Marmèche, 2004, 2005).

2. *Constraint management* allows designers to fine-tune their search for ideas. More specifically, we argue that designers' mental representations are shaped by a variety of constraints (e.g., Bonnardel, 2000), which govern their choices and decisions. We can identify several kinds of constraints (see, for instance, Bonnardel, 1999, 2000). Some constraints can be regarded as being either external to the designer, as in the case of *prescribed constraints* derived from a design brief or schedule of conditions, or internal to the designer (referred to here as *constructed constraints*), based on his or her previous experience and preferences. Other constraints (referred to here as *deduced constraints*) can be inferred by designers from an analysis of the current state of the design problem or from the implications of previously defined constraints. These different kinds of constraints can be linked to divergent thinking when they guide designers to look for ideas in a conceptual domain other than the one of the design product being conceived. They can also lead to convergent thinking, insofar as they help designers to assess ideas or solutions, and gradually delimit their search space until they reach a solution that is both new and meets all the various constraints.

According to the A-CM model, analogical thinking (as well as associations) and constraint management interact throughout the design process. Also, they influence other cognitive processes involved in design thinking, such as constructing mental representations, evaluating solutions, and adopting different perspectives.

This view is in line with Stokes (2007), who holds that implementing constraints reframes the problem and induces a new one for designers to solve. It is also consistent with Kelsey, Medeiros, Partlow, and Mumford (2014), who argue that constraints can generate creative solutions to problems. We believe that creative activities cannot take place (or at least only with considerable difficulty) unless constraints are taken into account, be they internal or external.

Our view is also in line with the GENErate and exPLORE (Geneplore) model (Finke, Ward, & Smith, 1992; Ward, Smith, & Finke, 1999; Ward, Smith, & Vaid, 1997). According to this model, there are two generic phases of creativity:

- a generative phase, in which mental representations, or *preinventive* structures, are constructed;
- an exploratory phase, in which these structures are explored in ways that lead to insights and discoveries.

These stages in the production of creative outcomes are seen as distinct, yet cyclical (Finke et al., 1992). Moreover, according to the *path of least resistance* described by Ward (1994), creativity is based on the activation of previous knowledge elements and on their recombination to generate new outputs.

Study of Professional Designers' Creative Design Thinking

Objective

This first study was conducted to analyze designers' perceptions of their own process of creative design thinking. More specifically, through interviews and questionnaires, we collected data about their perceptions of both their creativity and the stages in their design thinking process.

Method

Participants. We recruited a sample of 25 professional designers, comprising 19 men and 6 women, with an average of 14 years' experience in design.

Procedure. In line with the themes explored under the CREAPRO project, we applied a specific research protocol comprising a semi-structured interview and a questionnaire. The construction of the interview guide and the questionnaire was based on the multivariate approach to creativity devised by Lubart et al. (2003). This is a useful approach because it takes both the individual's characteristics and the environmental conditions into account. Therefore, we considered a variety of factors thought to influence the development of creative potential: cognitive (e.g., intelligence, knowledge), conative (e.g., personality, motivation), emotional, and environmental.

We conducted the *interviews* to analyze the perceptions that designers have of their own process of creative thinking and the stages in their creative process. These retrospective accounts of creative events in design provided by the designers themselves may not have been wholly reliable, but they did give us an overview of the designers' subjective perceptions of their own creative design process. The interviews were conducted individually, and their questions were divided into three phases: (1) general questions that allowed the designers to introduce themselves, say how long they had practiced their profession, and describe the training they had had; (2) questions about their creative productions; and (3) questions related to their process of design thinking—first a general description, then a more detailed one, focusing on what they considered to be their best creative production(s).

After the interviews, the professional designers were provided with a *questionnaire* that encouraged them to think about the factors described in the multivariate approach. This questionnaire consisted of a list of items liable to influence their creativity. In accordance with the multivariate approach, we selected four types of items: cognitive, conative, emotional, and environmental. Participants rated the importance that each item (and its sub-items) had for their creative process on a 7-point Likert-like scale.

- For the cognitive factors, the following sub-items were taken into account: general intelligence, ability to identify, define and redefine a problem, selective encoding, selective comparison, selective combination, divergent thinking, evaluating ideas, flexibility, domain-relevant knowledge, and convergent thinking.
- For the conative factors, the following sub-items were taken into consideration: perseverance, tolerance for ambiguity, individualism, risk taking, emotional stability, agreeableness, extraversion, consciousness, openness to new experiences (openness to dreaming, aesthetics, feelings, actions, ideas, and values), ability to take a step back, humbleness, self-criticism, motivation (intrinsic and extrinsic motivation, motivation to succeed, need for closure) and need to experience emotions.
- For the emotional factors, we probed emotional intelligence, emotional lucidity, ability to identify emotions, ability to regulate emotions, emotional granularity, alexithymia, emotional expressivity, affective intensity, emotional ambivalence, and emotional creativity.
- For the environmental factors, we explored constraints, physical environment, social environment, techniques, finances, and important life events.

For each item, we provided a clear definition of the vocabulary we used, so that all the designers would have a similar understanding of the terms we employed.

Results

After describing the results of the interviews and questionnaires, we summarize the major stages we identified in the process of design thinking as perceived by professional designers.

Specific stages based on interviews. To identify the different stages in the creative design process described by the designers during the interviews, we focused on their responses to the question "Are there any specific stages in your creative process? Could you name them?"

First of all, we should stress that even though all the designers were able to name the different stages in their creative design process, they produced nonlinear descriptions of their work. We recorded several comments like this one: "I do everything at the same time, I have to validate my work by all possible means, I constantly shift from drawings to mock-ups and from mock-ups back to drawings, because each time I find something unsatisfactory, and have to start all over again from the beginning."

We processed these data by (1) collecting the words used by the designers to describe and name the stages of their creative work and (2) placing them under different headings. Also, we took into account the order in which these stages were mentioned.

The overall results allowed us to create different headings corresponding to five stages in the process of design thinking (see Table 9.1): (1) *idea building*, based on documentation, as well as impregnation and analysis of the stimuli present in the environment; (2) *idea developing*, based on sketching and both divergent and convergent thinking; (3) *idea materializing* through mock-ups; (4) *realization or finalizing*, with regard to final constraints; and (5) *presentation to the relevant audience*, such as customers.

Factors Based on Questionnaires

To identify the main factors involved in the creative design process, we considered the responses given by all the designers to the questionnaires.

Overall results indicated that the designers considered the cognitive factors to be most important factors for their creative process (mean rating $= 5.36$; $SD = .75$). These were followed by conative factors ($M = 5.18$, $SD = .57$), environmental factors ($M = 4.95$; $SD = .96$), and emotional factors ($M = 4.13$, $SD = .88$).

We conducted a factor analysis to determine whether any regularities could be observed in the designers' ratings of specific items. We focused on cognitive items, given that they were considered to be the most important ones, and identified three factors:

Table 9.1 Stages mentioned by the professional designers during the interviews (the items frequently mentioned by the designers are highlighted in bold characters)

	Number of occurrences (25 interviews)
Stage 1: idea building	
Having an idea (*intuition, desire, motivation*)	**12**
Impregnation (*observing the environment, collecting experiences*)	5
Reflection (*analyzing, thinking about the problem*)	5
Building the brief (*thinking about constraints, listening to the customer*)	3
Documentation (*researching the customer, technology, benchmark*)	**12**
Incubation	1
Eureka	1
Stage 2: idea developing	
Sketching (*divergent thinking, drawing, shape searching, idea searching*)	**16**
Incubation	2
Experimenting (*tests, more accurate drawings*)	3
Convergent thinking (*searching for a concept, associating different constraints*)	**11**
Evaluation leading to choices (*being critical, choosing one or more solutions*)	3
Stage 3: idea materializing	
Mock-up (*exploring materials, visualizing details, validation, 3D*)	**13**
Rectifying (*trials, experimenting, correcting, iterations*)	6
Incubation	2
Stage 4: realization or finalizing	
Final constraints (*validating with customer, manufacturing constraints*)	10
Finalizing (*deadline, modifications, realization, prototype*)	**17**
Stage 5: presentation to audience	
Presentation (*delivery, final product, exposition, presentation to the customer*)	7

- Factor 1 (*problem-solving* profile), which grouped general intelligence, idea evaluation, and convergent thinking and showed negative loadings for divergent thinking and selective comparison;
- Factor 2 (*divergent* profile), which grouped divergent thinking, flexibility, and domain-relevant knowledge, and showed negative loadings for problem identification and selective comparison;
- Factor 3 (*problem-framing* profile), which grouped selective comparison and the ability to identify, define, and redefine a problem, and showed negative loading for domain-relevant knowledge.

General stages

The data we gathered allowed us also to identify three general stages that the professional designers regarded as particularly important in their process of creative design thinking:

- *Definition and redefinition of the creative problem* (corresponding to a *problem-framing* profile), with a mean rating of 6 on the 7-point scale. The designers explained that they initially consider the constraints provided in the design brief (or schedule of conditions), later supplementing them with others.
- *Openness to aesthetic dimensions, new experiences and new ideas* (corresponding to the *divergent* profile), with a mean rating of 6.43 on the 7-point scale. The designers explained that they draw ideas from other domains and that these ideas can be useful both for understanding the object to be designed and for dealing with the design problem at hand.
- *Self-assessment or reflexive evaluation* (partially corresponding to a *problem-solving* profile), with a mean rating of 6.11 on the 7-point scale. Here, designers evaluate their own creative productions. These self-assessments can be supplemented by external evaluations performed by other persons, especially people who are not involved in the design field.

These three main stages can be related to Botella et al. (2016)'s creative *macroprocesses*, in contrast to idea creation mechanisms, which can be regarded as *microprocesses*.

Discussion

We predicted that the professional designers in our sample would mention some stages in their creative design process more frequently than others, allowing us to identify the nature of their creative macroprocess. Three such stages were indeed identified: (1) definition and redefinition (or *problem framing*) of the creative problem; (2) openness to aesthetic dimensions, new experiences, and new ideas (possibly contributing to divergent thinking); and (3) reflexive evaluation (possibly contributing to convergent thinking). In addition, our analysis allowed us to highlight repetitive sequences that took place throughout the design process, similar to those described in the Geneplore model (Finke et al., 1992; Ward et al., 1999), where the generation of ideas is followed by their exploration.

During the *idea generation* stage, designers thought about how to build their own ideas, and chose one or more ideas they wished to develop. To this end, they evoked previous experiences in order to find sources of inspiration while thinking about the problem. This process was based on both analogies and the generation (or management) of constraints, in accordance with the A-CM model (Bonnardel, 2000, 2006). Our professional designers therefore engaged in the co-evolution of problem framing and problem solving described by Dorst and Cross (2001).

Idea development occurred when the designers externalized their ideas in the form of sketches, mock-ups or prototypes, in line with the exploration phase of the Geneplore model, characterized by experimenting and convergent thinking, rectifying, and finalizing. In this model, each sequence is repeated until the designer can move on to the next stage, although it is not definitive, as designers have often to go back to the previous stage, in order to rectify unsatisfactory features they have just discovered. The following stage begins with the validated form of the idea (with different levels of detail) developed in the previous stage (choice of an idea, validated mock-up, and final prototype). Our findings are also in line with some models of the creative process, such as those developed by Amabile (1996) and Mumford et al. (1991).

The *problem identification* or *problem construction* stage, which was described in different ways by designers participating in our study (e.g., desire for a specific theme, detection of a need in the environment, or just a sudden thought), gradually leads (possibly after iterations), either consciously or unconsciously, to *idea evocation*. The *documentation* stage in our description corresponded to the *preparation* stage in Amabile's model, sketching was part of *idea generation*, whereas mock-ups and finalizing mainly corresponded to the *validation* stage. Also we highlighted the importance of convergent thinking in design, which corresponds to the *synthesis* stage in McNeill et al. (1998)'s model. Finally, some designers participating in our study mentioned a *communication* stage, although not all of our participants considered it to be part of the creative process.

Study of Students' Creative Design Thinking

Objective and Hypotheses

This second study was conducted with design students to determine the influence of pedagogical methods on their design thinking process, these pedagogical methods focusing on either the evocation of ideas (in line with brainstorming) or the management of constraints (Bonnardel & Didier, 2016; Bonnardel, Mazon, & Wotjczuk, 2013). The students were exposed to one of the two types of design project instructions during short sessions and, then, they had to develop their design project.

We hypothesized that training students to produce ideas promotes divergent thinking and, by so doing, leads to more ideas. In contrast, training them to manage the constraints pertaining to the design project should promote the generation of yet more constraints, which frame the design problem and prompt convergent thinking. We expected therefore that these two types of training will have a differential influence on the number of ideas or constraints cited by design students in the early design phase. Moreover, they might also influence the nature of the constraints expressed by participants.

Two Design Project-Oriented Methods

In accordance with the above-mentioned models, and more especially the A-CM model (Bonnardel, 2000, 2006), we developed two project-oriented methods for design students. These shared similar bases, but were each intended to induce a different focus of attention (see Bonnardel & Didier, 2016; Bonnardel, Mazon, & Wotjczuk, 2013).

- The first method was intended to encourage designers to come up with creative ideas, in accordance with some of the stages identified in the previous study, and in line with the brainstorming method developed by Osborn (1963). In order to apply this method, called CQFD, participants had to follow four rules:
  - C for *No censorship*, where participants must avoid self-censorship and express all the ideas that come into their heads;
  - Q for *Quantity*, where participants have to write all these ideas down;
  - F for *Fantastic*, where participants have to express even the most crazy or extravagant ideas that spring to mind;
  - D for *Diversify*, where participants have to use different combinations of all the ideas they have expressed so far to find new ones.

- The second method was intended to encourage designers to take account of and manage the design problem's constraints, in accordance with other steps identified in the previous study. In order to apply this method, called CQHD, participants had again to follow four rules:
  - C for *Constraint*, where participants have to list all the constraints related to the problem that come into their heads;
  - Q for *Quantity*, where participants have to write all these constraints down;
  - H for *Hierarchization*, where participants have to rank the constraints they have listed;
  - D for *Diversify*, where participants have to use different combinations of all the constraints they have listed to find new ones.

Experiment

Participants. A total of 32 design students took part in this study. They were all in their first year of a design degree course. Because there was a limited number of students in each class in this specialized area, we asked students enrolled in two different design specialties to take part in our study: half of them had opted for a specialty in spatial design (SD), and half a specialty in product design (PD).

Procedure. Participants were divided into two groups, depending on the design method to which they were to be exposed (CQFD or CQHD). Each group consisted of 16 students (8 SD students and 8 PD students). To allocate the students to the CQFD or CQHD group, we asked their usual design teachers to construct pairs of students who were enrolled in the same design specialty and had similar mean scores on design tasks. One student in the pair was then assigned to the CQFD group, and the other to the CQHD group.

The procedure consisted of three phases: in the first and third phases, we tested the students to compare their performances before and after training, whereas in the second phase, all the students tackled the same design task (see below), but it was preceded by one or other of the two design methods.

Pre- and posttests. Each design student performed two versions of a test: one administered before the start of the experimental task (pretest) and the other after it (posttest). These pre- and posttests probed both divergent thinking (switching, morphing, and fluidity tests) and convergent thinking (e.g., selective combination test). For each of these tests, participants were provided with answer sheets and received both oral and written instructions for performing the tasks.

Creative design activity. The creative design activity was induced by a design brief provided to all the design students. This brief was defined in collaboration with design teachers, in order to suit students in both specialties, and it consisted of designing a new device to protect pedestrians crossing the road.

The students had 10 minutes to read and understand the schedule of conditions, after which they were asked to read a printed sheet setting out the rules corresponding to each experimental condition (CQFD or

CQHD). They were then given 20 minutes to write down all ideas and constraints that came to mind. Finally, they were allowed 90 minutes to represent their design project on A3 sheets and finalize sketches representing the design outcomes.

Data Analysis

Part of the data analysis was performed on the ideas and constraints written down by participants, and part was based on the evaluation of participants' creative productions by teachers specializing in creative activities (see Bonnardel & Didier, 2016). The data analysis allowed us to distinguish between ideas about the urban device to be designed and constraints related to the problem at hand:

- Ideas correspond to concepts that are more or less abstract, and define the *characteristics of the product to be designed*;
- Constraints correspond to *requirements* that the object to be designed has to satisfy. They can either be independent of the designer (*external* constraints) or be generated by the designer him- or herself (*internal* constraints). We made a distinction among three types of constraints (Bonnardel, 1999, 2000): (1) *prescribed* constraints (i.e., external constraints dependent upon the design brief); (2) *constructed* constraints (i.e., internal constraints arising from participants' own experiences or preferences); (3) and *deduced* constraints (inferred by designers from an analysis of the problem-solving process thus far or else from the consequences of other constraints).

Three analysts analyzed independently the data on the students' proposals. Their results were then compared, and any disagreements were discussed until a consensus was reached.

Results

After setting out the results for the characteristics of the students' design thinking process, focusing on the generation of ideas and constraints, we compared the participants' performances on the pre- and posttests.

Results on the students' design thinking. We ran analyses to determine whether (1) the activities performed by the SD and PD students differed significantly, (2) the type of training the students received affected the number of constraints and ideas they produced, and (3) the type of training affected the nature of the constraints that were listed.

- *Impact of design specialties*

Given that the SP and PD students were at the same level in their course and attended similar classes (albeit in different design specialties), we expected them to engage in similar design activities. Analyses of variance (ANOVAs) on the numbers of ideas and constraints produced by the SD and PD students failed to reveal any significant effect of design specialty on the numbers of ideas or constraints they expressed. Moreover, the ANOVAs showed no effect of design specialty on the generation of prescribed, deduced or constructed constraints.

- *Impact of type of training on the numbers of ideas and constraints*

Concerning the influence of type of training (CQFD vs. CQHD) on the numbers of ideas and constraints produced by the design students, results showed that training focusing either on the generation of ideas or on the management of constraints had a significant influence on both the number of ideas and the number of constraints that were produced. In accordance with our hypotheses, students who were exposed to the CQFD method produced more ideas on average ($M = 5.44$, $SD = 3.18$) than students who were exposed to the CQHD method ($M = 2.19$, $SD = 2.40$), $p = .003$. Conversely, the CQHD students listed more constraints ($M = 16.00$, $SD = 5.75$) on average than the CQFD students ($M = 9.25$, $SD = 3.69$), $p = .002$. Even so, regardless of training method, participants produced more constraints than ideas on average.

- *Impact of type of training on the nature of the constraints*

Concerning the influence of type of training on the nature of the constraints that were cited, an ANOVA revealed a significant effect of type of training on the total number of both prescribed constraints

(CQFD = 31,   CQHD = 79; $p < .01$),   and   deduced   constraints (CQFD = 37, CQHD = 84; $p < .01$), but there was no significant effect on the number of constructed constraints (CQFD = 80, CQHD = 93).

Results on the pre- and posttests. No significant effect of type of training was observed on posttest results (switching, morphing, fluidity and selective combinations). However, the pre- and posttests differed significantly on switching ($p < .01$), with higher mean scores after training than before in both the CQFD (24.33 *vs.* 21) and CQHD (25.33 *vs.* 21.33) conditions. A significant difference was also observed between the pre- and posttests for morphing ($p < .01$), with higher mean scores after training than before in both the CQFD (137 *vs.* 118) and CQHD (136.5 *vs.* 96) conditions. Both the switching and morphing tasks were measures of mental flexibility.

Discussion

The aim of this second study was to determine the influence of two types of project-oriented design methods (CQFD and CQHD) on students' design thinking process. In accordance with our hypotheses, results showed that the CQFD method prompted students in design to increase the number of ideas they generated, whereas the CQHD method stimulated them to list more constraints for dealing with the design project. Thus, although participants were only briefly exposed to these two methods, the latter had a differential effect on their creative design thinking process. Students who were exposed to the CQHD project-oriented method seemed to be more engaged in framing (defining and redefining) the design problem at hand than students who were trained in the CQFD method, who produced more ideas and seemed more engaged in solving the design problem. The design project-oriented methods provided to the students in our study therefore appeared to modify the way they tackled the design problem. Our results revealed also that participants were able to adapt their procedures to the rules imposed by the training task, underscoring the flexibility of the creative process.

Conclusion

Our general objective in this chapter was to contribute to a better understanding of the process of design thinking. We began by describing the characteristics of design activities, as well as descriptive models of creativity and design thinking. Then we extended these descriptions with the results of a study conducted with professional designers to probe their perceptions of their own creative design thinking process. Several stages and factors were highlighted, but we focused on cognitive factors, as these were the factors that the professional designers deemed to be most important. We were able to identify three general stages: (1) definition and redefinition of the creative problem (corresponding to a *problem-framing* profile); (2) openness to aesthetic dimensions, new experiences and new ideas (corresponding to a *divergent* profile); and (3) self-assessment or reflexive evaluation (partially corresponding to a *problem-solving* profile). In view of the importance of these three stages in design thinking, and the nature of the difficulties that can be encountered along the way, it might be possible to support designers during these three stages through human–computer interaction and even human–computer cooperation (Bonnardel & Zenasni, 2010; Burkhardt & Lubart, 2010; Hewett, 2005). In particular, designers could be provided with computational systems such as TRENDS, which provides users with sources of inspiration in the form of images, or SKIPPI, where the sources of inspiration are provided in word form (Bonnardel & Bouchard, 2014; Bonnardel & Zenasni, 2010). The use of a critiquing expert system could also favor the evaluation of design solutions (see, for instance, Bonnardel & Zenasni, 2010).

Next, we described a second study that we conducted with design students to analyze the influence on their design thinking process of two methods encouraging them to focus on either idea generation (in line with divergent thinking) or constraint management (in line with both problem-framing and convergent thinking). Results of this second study revealed differential effects on students' creative design thinking process, which were confirmed by the assessment of their creative productions (Bonnardel & Didier, 2016). These results could inform the development of new teaching methods favoring the creative design process.

These two studies complemented each other, insofar as they allowed us to analyze the design thinking processes of both professional designers and design students. These studies yielded results that should contribute to the emergence of new educational and/or computational modalities that favor the creative design process (see, for instance, Bonnardel, 2016).

Acknowledgements We would like to thank all the professional designers who took part in the first study, as well as the design students at the Perrin and Diderot high schools in Marseilles, and their teachers René Ragueb, Cathy Bourgoin and Véronique Billaud, for their contribution to the second study described in this chapter. This research was performed under the CREAPRO French National Research Agency contract (grant no. ANR-08-CREA-038).

References

Amabile, T. M. (1996). *Creativity in context*. Boulder, CO: Westview Press.

Asimov, M. (1962). *Introduction to design*. Englewood Cliffs, NJ: Prentice-Hall.

Bonnardel, N. (1999). L'évaluation reflexive dans la dynamique de l'activité du concepteur. In J. Perrin (Ed.), *Pilotage et évaluation des activités de conception* (pp. 87–105). Paris: L'Harmattan.

Bonnardel, N. (2000). Towards understanding and supporting creativity in design: Analogies in a constrained cognitive environment. *Knowledge-Based Systems, 13,* 505–513.

Bonnardel, N. (2006). *Créativité et conception: Approches cognitives et ergomiques* [Creativity and design: Cognitive and ergonomic approaches]. Brussels: De Boeck.

Bonnardel, N. (2009). Activités de conception et créativité: De l'analyse des facteurs cognitifs à l'assistance aux activités de conception créatives. *Le Travail Humain, 72,* 5–22.

Bonnardel, N. (2012). Designing future products: What difficulties do designers encounter and how can their creative process be supported? *Work, A Journal of Prevention, Assessment & Rehabilitation, 41,* 5296–5303.

Bonnardel, N. (2016). Propositions de méthodes d'analyse et de modalités d'assistances pédagogique et informatique aux activités créatives. Illustrations dans le domaine du design. In I. Capron-Puozzo (Ed.), *La créativité en éducation et en formation. Perspectives théoriques et pratiques* (pp. 167–180). Paris: Albin Michel.

Bonnardel, N., & Bouchard, C. (2014). Design, ergonomie et IHM: Etudes complémentaires pour favoriser les activités de conception créatives. In N. Couture, C. Bastien, & T. Dorta (Eds.), *Quelle articulation pour la co-conception de l'interaction? Proceedings of the international conference 2014 Ergonomie et Informatique Avancée Conference - ErgoIA'2014* (pp. 33–40). Toulouse and New York: ACM Press.

Bonnardel, N., & Didier, J. (2016). Enhancing creativity in the educational design context: An exploration of the effects of design project-oriented methods on students' evocation processes and creative output. *Journal of Cognitive Education and Psychology, 15*(1), 80–101.

Bonnardel, N., Didierjean, A., & Marmèche, E. (2003). Analogie et résolution de problèmes. In C. Tijus (Ed.), *Métaphores et analogies* (pp. 115–149). Paris: Hermès.

Bonnardel, N., Forens, M., & Lefevre, M. (2016). Enhancing collective creative design: An exploratory study on the influence of static and dynamic personas in a virtual environment. *Design Journal, 19*(2), 189–203.

Bonnardel, N., & Marmèche, E. (2004). Evocation processes by novice and expert designers: Towards stimulating analogical thinking. *Creativity and Innovation Management, 13*(3), 176–186.

Bonnardel, N., & Marmèche, E. (2005). Towards supporting evocation process in creative design: A cognitive approach. *International Journal of Human-Computer Studies, 63,* 442–435.

Bonnardel, N., Mazon, S., & Wojtczuk, A. (2013). Impact of project-oriented educational methods on creative design. In *Proceedings of the 31st European Conference on Cognitive Ergonomics—ECCE 2013*, Toulouse, France, article no. 6. New York: ACM Press.

Bonnardel, N., & Zenasni, F. (2010). The impact of technology on creativity in design: An enhancement? *Creativity and Innovation Management, 19*(2), 180–191.

Botella, M., Nelson, J., & Zenasni, F. (2016). Les macro et microprocessus créatifs. In I. Capron-Puozzo (Ed.), *La créativité en éducation et en formation. Perspectives théoriques et pratiques* (pp. 31–44). Paris: Albin Michel.

Burkhardt, J.-M., & Lubart, T. (2010). Creativity in the age of emerging technology. *Creativity and Innovation Management, 19,* 160–166.

Chrysikou, E. G., & Weisberg, R. W. (2005). Following the wrong footsteps: Fixation effects of pictorial examples in a design problem-solving task. *Journal of Experimental Psychology. Learning, Memory, and Cognition, 31,* 1134–1148.

Csíkszentmihályi, M. (1996). *Creativity: Flow and the psychology of discovery and invention.* New York: HarperCollins.

Dewey, J. (1934). *Art as experience.* New York: Penguin.

Dorst, K., & Cross, N. (2001). Creativity in the design process: Co-evolution of problem-solution. *Design Studies, 22,* 425–437.

Eastman, C. M. (1969). Cognitive processes and ill-defined problems: A case study from design. In *Proceedings of the 1st International Joint Conference on Artificial Intelligence* (pp. 669–690). Washington, DC.

Finke, R. A., Ward, T. B., & Smith, S. M. (1992). *Creative cognition: Theory, research, and applications.* Cambridge, MA: MIT Press.

Gero, J. S. (1998). Towards a model of designing which includes its situatedness. In H. Grabowski, S. Rude, & G. Grein (Eds.), *Universal design theory* (pp. 47–56). Aachen: Shaker Verlag.

Glăveanu, V. P. (2012). Creativity and folk art: A study of creative action in traditional craft. *Psychology of Aesthetics, Creativity, and the Arts, 7*(2), 140–154.

Glăveanu, V. P., Lubart, T., Bonnardel, N., Botella, M., de Biaisi, P.-M., Desainte-Catherine, M., … & Zenasni, F. (2013). Creativity as action: Findings from five creative domains. *Frontiers in Educational Psychology, 4,* 176.

Hayes-Roth, B., & Hayes-Roth, F. (1979). A cognitive model of planning. *Cognitive Science, 3,* 275–310.

Hewett, T. T. (2005). Informing the design of computer-based environments to support creativity. *International Journal of Human-Computer Studies, 63,* 383–405.

Jansson, D. G., & Smith, S. M. (1991). Design fixation. *Design Studies, 12,* 3–11.

Kelsey, E., Medeiros, P., Partlow, J. P., & Mumford, M. D. (2014). Not too much, not too little: The influence of constraints on creative problem solving. *Psychology of Aesthetics, Creativity, and the Arts, 8,* 198–210.

Lubart, T., Mouchiroud, C., Tordjman, S., & Zenasni, F. (2003). *Psychologie de la créativité.* Paris: Armand Colin.

McNeill, T., Gero, J. S., & Warren, J. (1998). Understanding conceptual electronic design using protocol analysis. *Research in Engineering Design, 10,* 129–140.

Mumford, M. D. (2003). Where have we been, when are we going? Taking stock in creativity research. *Creativity Research Journal, 15*(2–3), 107–120.

Mumford, M. D., Mobley, M. I., Reiter-Palmon, R., Uhlman, C. E., & Doares, L. M. (1991). Process analytic models of creative capacities. *Creativity Research Journal, 4,* 91–122.

Osborn, A. F. (1963). *Applied imagination: Principles and procedures of creativity thinking*. New York: Charles Scribner's Sons.

Rittel, H., & Webber, M. M. (1984). Planning problems are wicked problems. In N. Cross (Ed.), *Developments in design methodology* (pp. 135–144). New York: Wiley.

Schön, D. A. (1983). *The reflective practitioner: How professionals think in action*. New York: Basic Books.

Simon, H. A. (1973). The structure of ill-structured problems. *Artificial Intelligence, 4,* 181–201.

Simon, H. A. (1995). Problem forming, problem finding and problem solving in design. In A. Collen & W. Gasparski (Eds.), *Design & systems* (pp. 245–257). New Brunswick: Transaction Publishers.

Stokes, D. (2007). Incubated cognition and creativity. *Journal of Consciousness Studies, 14,* 83–100.

Sternberg, R. J., & Lubart, T. (1999). The concept of creativity: Prospects and paradigms. In R. J. Sternberg (Ed.), *Handbook of creativity* (pp. 3–15). New York: Cambridge University Press.

Tan, S., & Melles, G. (2010). An activity theory focused case study of graphic designers' tool-mediated activities during the conceptual design phase. *Design Studies, 31,* 461–478.

Treffinger, D. J. (1995). Creative problem solving: Overview and educational implications. *Educational Psychology Review, 7,* 301–311.

Visser, W. (1994). Organisation of design activities: Opportunistic, with hierarchical episodes. *Interacting with Computers, 6,* 235–238.

Wallas, G. (1926). *The art of thought*. New York: Harcourt-Brace.

Ward, T. B. (1994). Structured imagination: The role of category structure in exemplar generation. *Cognitive Psychology, 27,* 1–40.

Ward, T. B., Smith, S. M., & Vaid, J. (1997). Conceptual structures and processes in creative thought. In T. B. Ward, S. M. Smith, & J. Vaid (Eds.), *Creative thought: An investigation of conceptual structures and processes* (pp. 1–27). Washington, DC: American Psychological Association.

Ward, T. B., Smith, S. M., & Finke, R. A. (1999). Cognition. In R. Sternberg (Ed.), *Handbook of creativity* (pp. 189–212). Cambridge: Cambridge University Press.

Zeisel, J. (1981). *Inquiry by design: Tools for environmental behavior research*. Cambridge, MA: Cambridge University Press.

10

Creative Thinking in Music

Baptiste Barbot and Peter R. Webster

Introduction

Musical creativity of eminent musicians, primarily composers, has received attention for centuries. Its systematic study, however, has begun only recently. The increased interest concerning musical creativity as a

The preparation of this chapter was partly supported by grant RFP-15-05 from the Imagination Institute (www.imagination-institute.org), funded by the John Templeton Foundation. The opinions expressed in this publication are those of the authors and do not necessarily reflect the view of the Imagination Institute or the John Templeton Foundation.

B. Barbot (✉)
Department of Psychology, Pace University, New York, NY, USA
e-mail: bbarbot@pace.edu

B. Barbot
Child Study Center, Yale University, New Haven, CT, USA

P. R. Webster
Thornton School of Music, University of Southern California, Los Angeles, CA, USA

© The Author(s) 2018   **255**
T. Lubart (ed.), *The Creative Process*, Palgrave Studies in Creativity and Culture,
https://doi.org/10.1057/978-1-137-50563-7_10

scientific object of study corresponds with the "democratization" of the concept of musical creativity less than half a century ago, then understood as a facet of the human creative potential that can be nurtured and studied in the "general population" (e.g., Barbot & Lubart, 2012). These advances in the conceptualization of musical creativity have led to substantial progress in the understanding of various aspects of creative behaviors in music, including the creative thinking[1] process.

After outlining an inclusive approach to understanding individual differences in musical creativity outcomes, and providing a snapshot of the literature on the empirical study of musical creativity, this chapter presents several lines of work by authors who have studied the creative thinking process in music. We highlight a number of factors that impact this process, including the general approach to composition (or improvisation),[2] age, formal musical education and expertise, as well as the effect of technology and material conditions. These factors are then illustrated through the presentation of an empirical study exploring the relationship between the musical processes and products of novice adolescents' composers. We conclude by outlining the critical role of "intention," and therefore the role of the composer's agency, as a fundamental and primary requirement for a successful musical creative process, as well as the critical role of creative opportunities for this process to take place.

---

[1]We have chosen to use the term "creative thinking" when addressing those processes associated with the behaviors of music creation, in this case, by children and adolescents in a composition context. The word "creativity" is seen as a more general term to encompass inventiveness in a more holistic context.

[2]The majority of the literature reviewed here is based on the musical experience of composition. Improvisatory behavior is certainly a strong part of the creative process of composition as a product intent but the focus of this chapter is primarily on composition and not on the processes of improvisation per se. Related studies of creative thinking in improvisation, music listening or music performance with children and adolescents warrant continued study as separate topics.

We Are All Musically Creative (to Some Degree)

With the exception of some early descriptive and qualitative work (e.g., Coleman, 1922; Moorhead & Pond, 1941–1944), the systematic study of creative thinking in music is a rather recent endeavor of the past five decades. Although musical creativity has been studied in various fields, including sociology, cognitive science, and artificial intelligence, the most extensive literature on this topic is in the field of music education. In this specific field, the objective is usually to better understand the nature of creative thinking and behavior in music, in order to promote and incorporate these characteristics in educational programs. Interestingly, research in music education is often colored by the legacy of psychological theories and models such as those by Csikszentmihalyi (1988), Gardner (1983), or Guilford (1967), and by many theories from the subfield of the psychology of music (e.g., Sloboda, 1985, 1988; Teplov & Deprun, 1966). Yet, the overall topic of musical creativity has only been studied in a limited way from a psychological perspective (e.g., Leman, 1999). Given that the majority of empirical knowledge on creative thinking in music is based on music education research, most of what we know on the topic is derived from "healthy" children and/or adults with some musical training.

However, it is increasingly acknowledged that musical creativity is an expression of human creative potential that can be nurtured and studied in the "general population" (e.g., Barbot & Lubart, 2012). This is becoming particularly true given the disappearance of the distinction between musical outcomes produced by formally educated experts versus informally trained amateurs. It is indeed well established that creativity (regardless of content domain) is not an ability solely demonstrated by individuals with advanced specific domain knowledge and skills (although this knowledge and skill may contribute to the quality of creative outputs): there are individual differences in creative potential and resulting creative outputs. Recent "hybrid" models of creative potential outline the contribution of domain-general, domain-specific, and task-specific resources in any creative endeavor (e.g., Barbot, Lubart, & Besançon, 2016; Kaufman & Baer, 2004).

By extension, an individual does not have only one creative potential,[3] but as many potentials for creativity than outlets in which creativity can be expressed (Barbot, Besançon, & Lubart, 2015; Barbot et al., 2013; Lubart, Zenasni, & Barbot, 2013). Accordingly, the quality of creative outcomes in a specific domain or task (such as in musical composition, within a particular musical genre) will depend and the "quality of fit" between an individual's resources (i.e., profile of domain-general, domain-specific, and task-specific resources) and a subtle mixture of task-specific demands (Barbot et al., 2016).

This modern view of the human creative potential is rooted in the extension of the concept of creativity in the 1950s (e.g., Guilford, 1950), thus far, mainly associated with the study of exceptional individuals, and then re-interpreted as a psychological dimension that is normally distributed in the general population (and not exclusively to those with domain-specific training). Interestingly, in this generalization of the concept of creativity, music has been seen by some to be the exception to the rule. Adorno (1976), for example, deplored the specific "status" of music which, in the common sense, requires a "special gift," contrary to other creative outlets such as poetry or painting for which such exceptional abilities do not seem required in order to lead to outstanding creative outputs. This assumption is clearly contradicted by contributions in the fields of music education, musicology, and psychology of music, that have amply demonstrated that there is a "spontaneous" musical activity as well as abilities of perception and understanding of music from the very early years of the life (e.g., Barrett & Tafuri, 2012; Bigand & Poulin-Charronnat, 2006; Kogan, 1994).

With the advent of music gamification, new music technology, musical genre, and the accessibility of these new tools that do not necessarily rely on classic music notation, technical mastery of an instrument or "established" creative processes, it becomes apparent that music is an outlet of creative expression for many people that have limited or no formal musical training, sometimes attaining outstanding recognition in

---

[3]This view would suggest that creativity is only a generalized ability ("g-factor view of creativity"; Barbot & Tinio, 2015).

their field (Tobias, 2012). In all, it is a reasonable assumption to interpret musical creativity as a latent potential in all individuals (whether musically trained or not), leading to outputs of varying quality on the creativity "continuum," depending on the fit between content area requirements, individual-level resources, environmental influence, time and place, and more importantly, whether individuals have the desire and opportunity to express their potential (e.g., Barbot, Tan, & Grigorenko, 2013).

The Systematic Study of the Creative Process in Music

Studies of musical creativity can be broadly distinguished within a *Focus* × *Approach* × *Method* framework. With regard to their *Focus*, studies differ according to whether they are focusing on the product and/or the process in improvisation and/or musical composition. The difficulty to distinguish improvisation and composition has led to a long-standing debate which is becoming less relevant as new forms of musical expression and creation have appeared (Burnard, 1999, 2012; Hargreaves, Miell, & MacDonald, 2012; Webster, 2016a). A related argument that has fueled this debate was that the distinction between process and product might be unclear in the case of improvisation because the process is simultaneously the product (e.g., Eisenberg & Thompson, 2003). Other areas of focus have included less commonly studied topics such as creative musical listening sometimes considered as an important part of the creative experience in music (e.g., Dunn, 1997; Webster, 1987).

Regarding the *Approach* perspective, studies of musical creativity are also distinguished according to the theoretical and conceptual anchoring guiding the research process (i.e., theoretical lens, hypotheses formulation, etc.). These approaches fall within three main traditions in the study of musical creativity: individualistic, sociocultural, and "hybrid" (Sawyer, 2012; Webster, 2016a). Individualistic approaches attribute musical outcomes to personal characteristics within a usually homogeneous (Western) cultural context, and rely mainly on quantitative methods. Sociocultural approaches are usually less focused on process and product and more related to cross-cultural issues or musicianship roles, and they rely mainly on qualitative methods. In a thorough review of the literature on student composition in recent years, Webster (2016a) noticed the emergence of a "hybrid" approach that combines the traditional "individualistic" approach with a (re)emerging "sociocultural" approach, by considering both individual and sociocultural resources in the creative process (e.g., Folkestad, 2012).

Correspondingly, studies of musical creativity can be broadly distinguished by their *Method* (often guided by the study's main *focus* and underlying *approach*). Multiple methods for studying the creative process in music have been used including think-aloud or retrospective semi-structured interviews of people engaged in creative activities in music (e.g., Barrett, 2000; Glăveanu et al., 2013), analyses of manuscripts (Sloboda, 1988), or recorded musical streams at various stages of the creative process (e.g., Folkestad, 2012; Hickey, 1997; Seddon & O'Neill, 2003), the systematic observation of subjects composing or improvising music (e.g., Brand, 2000; Kratus, 1989; McPherson, 2000), or a combination of these approaches (e.g., Barbot & Lubart, 2012). Noticeable in this past decade has been the increasing body of neuroimaging studies (e.g., Bashwiner, Wertz, Flores, & Jung, 2016) examining the neural correlates of musical creativity, especially in music improvisation (for a review, see Beaty, 2015).

From Macro- to Microprocess in Musical Composition: A Focus on Compositional Problem Discovery and Musical Exploration of Novice Composers

Through their diversity of *Focus*, *Approach* and *Method*, decades of music creativity studies have clarified the creative process in music under complementary angles and perspectives. It is now becoming so accessible for young people to make and distribute the products of their musical thinking through musical composition that this type of musical activity is increasingly acknowledged as a fundamental part of what they naturally do with music (Kaschub & Smith, 2009; Webster, 2016a). Illustrating the increasing interest in the study of both compositional products and process in students, the National Core Arts Standards (2014; http://www.nationalartsstandards.org/) have incorporated a number of competencies that students should develop with regard to music composition including exploration, selection of sound sources in composition (and justification of choices made), and showing original, unique, and expressive ideas.

Similar to many other domains of creative endeavor, a number of studies of the creative thinking process in music have relied on Wallas's (1926) model and considered musical creativity as a multiphase problem-solving process including *preparation, incubation, illumination,* and *verification* (e.g., Feinberg, 1973; Webster, 1987). Wallas's model and its applications in the musical field consider the creative process as a sequential process within a rather large "time-scale" (i.e., from problem "definition" to "solution"). In this respect, this view of the process mainly focuses on "macro-level" events that delineate the creative process. One such macro-level process in musical composition is "problem discovery," considered as the first step of the creative process (similar to Wallas's preparation). This process, defined by Getzels (1964) and extended by Getzels and Csikszentmihalyi (1976), involves the discovery and formulation of the creative problem itself (involving micro-level processes such as exploration and selection of useful elements for the realization of a creative product).

Another line of work has mainly focused on these micro-level processes and events that constitute each of the macro-level processes (such as exploratory behaviors that are part of the "problem discovery" or "preparation" stage). Indeed, many studies on creativity in general and musical creativity, in particular, stressed the importance of the process of exploration, without which creativity cannot be expressed (e.g., Barbot & Lubart, 2012; Sternberg, 1988; Webster, 1996).

In conceptualizing specifically the creative thinking process in music and in embracing both macro- and micro-level considerations, Webster (2002) refined his original model (1987) by expanding central processes to include exploration with what he called "primitive gesturals." Also included were stages related to time away (possible incubation), working through (revising, editing, forming new ideas), and verification (rehearsal and polishing). The process was seen as quite flexible—aided by both convergent and divergent thinking and enabled by various skills and conditions (personal and social).

Bamberger (1977, 2013) used a computer-based composition method well before current technology to study the decision and exploration processes in the composition of melodies. Her analyses revealed a distinction between "prudent" and "impulsive" explorations that reflects individual differences in music representation (i.e., ability to "think in sounds"). Clear relationships were demonstrated between problem discovery behaviors (in particular, the quantity and originality of explorations) and the creative "quality" of resulting products in visual arts (Getzels & Csikszentmihalyi, 1976), but also in music (Barbot & Lubart, 2012; Brinkman, 1999; Webster & Hickey, 1995).

In a similar line, Kratus's (1989) nonhierarchical process model includes the processes of exploration (sound experimentation with available instruments), development (variations from an initial musical segment), repetition (in which the individual replays exactly the same musical segment that was produced during a process of exploration or development) and silence (no observable activity, which could refer to a form of micro-level incubation process, as described by Wallas). Using recording and observation of children of different ages during their compositional process, Kratus's paradigm was used to distinguish quantitative differences in the use of micro compositional processes. His conclusions indicate that, as they get older, children use less the process of exploration and more the process of development and repetition (Kratus, 1989). In another study, Kratus (1990) suggests that the most creative children use a wide variety of exploration strategies, development and repetition, and tend to converge toward a final solution relatively early in their creative process. These results, and insights from several other studies focused on the creative process, illustrate the relative dependence of macro- and micro-level processes and their impact on the resulting musical product.

Relationship Between Macro- and Micro-level Creative Processes in Music: The Impact of Age, Experience, and Material Conditions

Consider the case of a child who is inexperienced in composition and/ or technical mastery of a given instrument, engaging in the composition of a short musical piece that incorporates the typical "ingredients" of a musical object (e.g., identifiable structure, melodic contour, rhythmic motives). How, without musical training, will this child end up with a new and original product? Will his or her spontaneous or even more refined thinking processes impact the creative quality of the musical product? These questions were directly or indirectly explored in a small body of studies focusing on the compositional process of novice children and adolescents (e.g., Barbot & Lubart, 2012; Seddon & O'Neill, 2003). This literature has often identified various "naïve" compositional styles or approaches that intriguingly resemble the more expert compositional processes of eminent musicians (e.g., Folkestad, Hargreaves, & Lindström, 1998).

Based on an analysis of scores produced at various stages of the creative process, Sloboda (1985) argued that Beethoven had a "horizontal" compositional strategy (i.e., producing an instrumental part completely before working on another part), whereas Mozart had a "vertical" compositional strategy (i.e., simultaneously developing several instrumental parts, moving from one section to another). It may be agued that vertical composition strategies are comprehensive, holistic approaches (perhaps showing that the individual has already conceptualized the individual parts), whereas the horizontal strategies are more analytic, because individual work on each part proceeds sequentially. Although Sloboda's analysis indicates that both strategies may be associated with exceptional potential (as illustrated by Mozart and Beethoven's work), Folkestad and colleagues (1998) have outlined that both strategies are also naturally employed by naïve composers.

However, there are also a number of studies (e.g., Davidson & Welsh, 1988; Velia, Irvin, Berendt, & Ramirez, 1999) suggesting that expert music composers tend to write their musical pieces entirely after having fully imagined them (reflecting a more holistic or "vertical" approach), whereas novices begin to compose without a clear idea or representation of the final production (hence, leading to a more "horizontal" approach). This suggests that there might be an influence of the level of expertise with the general approach to composition (i.e., horizontal vs. vertical strategies), but also on the microprocesses involved in musical production, such as instrumental exploration. In their experiment centered on computer-based music composition of adolescents, Seddon and O'Neil (2003) found that adolescents with formal musical experience tended to explore fewer instruments than adolescents without formal musical experience. A possible explanation is that musical training provides skills that allow developing musical ideas outside the compositional set-up. That is, musical training may somehow help internalizing the instrumental exploratory process using anticipatory mental images. In line with Seddon and O'Neil (2003) and developmental studies by Kratus (1989), Hewitt (2009) found that (1) older students explored less and worked faster than younger students and that (2) students with prior music training explored also less than students with limited training.

In addition to a composer's personal resources illustrated above, it is likely that external resources and the environment, in which the creative process takes place, influence also that process. One aspect of this environment is technology, and other material conditions for composition. These conditions may orient or "encourage" the compositional process in a particular approach. In studies that have called upon the use of a sequencer or a music notation program as the recording platforms to produce musical objects (Barbot & Lubart, 2012; Folkestad et al., 1998; Hickey, 1997; Webster, 2012), the particular use of these technologies, general instructions, and parameters for composition could encourage a more sequential approach (i.e., leading to horizontal strategy). In "natural" conditions, the particular set-up for composition may greatly vary from musical genre to subgenre, and from composer to composer. As such, the creative process in music may be somewhat "content"-specific: it may not only vary as a function of the type of musical object to be created (improvisation, composition, sound object, etc.) but also as a function of the musical genre or subgenre involving a typical set-up required for music production, and differences in the general workflow (e.g., use of sequencers vs. traditional music notation, nature of the instrument employed, reliance on live, vs. postproduction processing). For example, contrary to the creative process engaged in classical music composition, many electro-acoustic music composers rely heavily on "technology that permits the integration of effects, insertion, mixing, dividing, synthesizing, modulating and multiplying segments, compressing and decompressing, cutting and reorganizing" (Glăveanu et al., 2013, p. 11). However, the effect of technology and material conditions for musical creativity remains a topic only beginning to be studied (Nevels, 2013; Webster, 2011).

Impact of the General Approach to Composing on Microprocesses: An Empirical Study

The general approach to composition per se might in turn impact the microprocesses involved in the creation of a musical piece. For instance, horizontal and vertical strategies might structure the phases of instrumental exploration, development, repetition, and silence (see Kratus, 1989), at different moments of the creative process. Horizontal strategies might lead to more instrumental explorations as the musical piece builds up sequentially (e.g., once a first instrumental part is composed, the subsequent parts are being developed, which might require new phases of exploration and reconsideration of previous recordings).

This hypothesis was explored in an unpublished study on creative processes and products in musical composition, in relation to dimensions of the self including domain-specific self-concepts and general self-esteem (Harlow & Barbot, 2015). In this study, the Musical Expression Test (MET; Barbot & Lubart, 2012) was administered to 102 adolescents with limited to no musical training background. The MET is a multimethod assessment technique which involves the systematic observation of musical exploratory behaviors as well as a product-based assessment of short musical piece composed using computer-based recordings and a set of diverse and playful instruments. During the composition process of each adolescent, exploratory behavior as well as indicators of compositional elaboration were systematically observed, coded and quantified. These indicators were used to identify ultimately typical "styles" in the compositional approach of adolescents.

Four typical styles were identified: one (relatively rare) could be classified as a holistic more "vertical approach," whereas the three others could be defined as "horizontal approaches" to composition. Specifically, the "vertical" style consisted mainly of a long problem-discovery phase (long period within which adolescents explored and set the main parameters of their composition by alternating from one section to the next to build a coherent ensemble or to coordinate "response" games between instrumental sections). This phase was followed by a series of recording all composition elements with limited repetition or revision between recordings (suggesting that participants stayed with their initial idea).

In contrast, an "improvisational" approach consisted of a limited or unexisting problem-discovery phase prior to recording the first sequence, followed by a short exploration phase between subsequent recordings, but numerous elaboration events (e.g., mixing, re-arranging) suggesting a constant re-interpretation of the musical object.

An "academic" style emerged, consisting of numerous repetitions of musical segments of rather unoriginal instrument uses before recording them, in a sequential approach, and with limited revision.

Finally, the "sound-designer" style was characterized by a perfectionist approach with a focus on "in-depth" exploration leading to original instrument uses and numerous re-execution/revisions of segments in recording phases (until results appeared satisfactory).

All compositions resulting from the MET were scored with satisfactory agreement by four quasi-experts using the Consensual Assessment Technique (Amabile, 1982). Although the prevalent use of a horizontal approach in this study is in line with the typical novice approach to musical composition described in the literature (e.g., Davidson & Welsh, 1988), analyses did not reveal any sizable differences between the four creative styles outlined here and the creativity of the resulting musical outputs, as assessed by the raters. In other words, different pathways, approaches, and microprocesses involved in the creative work in musical composition may all result in a similar range of outcomes (i.e., from low to high creativity).

However, the four styles were interestingly associated with the various dimensions of the "self" under investigation. For example, the "academic" style related to the highest creative self-efficacy and the "vertical" style the lowest, whereas the "improvisational" style was associated with the highest social self-esteem. These group differences reflected moderate associations between specific microprocesses (exploration, compositional elaboration, etc.) that were typical of each compositional style, and the various dimensions of self.

In sum, compositional strategies and general approaches to composition may not only reflect the various level of formal training received by the composer, but also some more intrinsic characteristics such as their cognitive style and identity-related dimensions. In this line, Seddon and O'Neil (2003)'s interpretation of the reduced exploratory behaviors in adolescents with formal musical training (see also results of Hewitt, 2009) was that they were implicitly encouraged to confirm their "musician" status and to maintain that image by only producing musical segments that are "appropriate." Together, these insights from novice adolescents suggest that individual resources (including knowledge, personality, and self-characteristics) orient the general approach to creativity, and in turn, impact both macro- and micro-level creative process. This creative process, however, seems independent from the relative "creative quality" of the musical output resulting from that process. Of course, it should be stressed that these generalizations require much more rigorous investigation under varying compositional settings.

Concluding Remarks: There Is no Creative Process in Music Without Agency and Opportunities for Musical Expression

As introduced above, while it is acknowledged that there is a (latent) musical creativity in every one of us, it is surprising how so few people achieve actually their potential in this area of creative endeavor. One of the main raisons for this is that people need the *creative intent* to achieve their potential as well as the *opportunity* to express their potential (e.g., Barbot et al., 2016). This points to two promising directions for a better understanding of the creative process in music, including (1) a better understanding of an individual's agency (Wiggins, 2015) in the creative process (including the notion of creative intent, desire, and motivation) and (2) the importance of providing children with regular creative music experiences including music listening (Barrett & Webster, 2014), that could help them formulate a musical "discourse" (Folkestad, 2012), equally important in understanding the creative work in music.

Beyond a number of factors that have been outlined in this chapter and elsewhere (cognitive styles/general approach to music composition, expertise and technical resources involved in the creative process), a key factor that remains rather unexplored to date is a person's agency (i.e., intention motivation or desire) in the decision to express herself/himself creatively in music. Although these factors seem prototypical of the "individualistic" approach to understanding creativity criticized by many (e.g., Sawyer, 2012), "agency" embraces a set of factors that transcends both individualistic and sociocultural interpretations of creativity. Indeed, whereas both interpretations tend to treat the composer as a passive agent in the creative work—an agent that is more or less successful in his or her creative enterprise depending on a number of dispositional and circumstantial factors highlighted by the individualistic and sociocultural approaches, respectively—there is an increasing acknowledgment of the role of the composer as active agent in the creative process through conceptualization of the notion of "product intent" (Webster, 1987, 2002) or the critical issue of intentionality in the composition process and its implication for instruction (e.g., Kratus, 2012). In particular, "creative achievement for children and adults is driven certainly by personal characteristics such as innate talent and personality, but more importantly by continued opportunities to compose, improvise, perform music of others with creative intention, and to listen to music creatively" (Webster, 2016b, p. 27).

Considering the gamification of musical creativity initiated by early music video games and music trackers in the 1980s and the more recent explosion of modern music games and entry-level music production platforms (e.g., guitar hero, garage band), it seems that there is increasing involvement of less formally music-educated people in musical activity, which, as outlined earlier, sometimes reach the creative attainment of formally educated experts. This suggests that music gamification may represent an area of opportunity for musical expression and commitment of amateur musicians. However, the most critical area of opportunity in this respect is the need for more numerous and regular creative music experiences that happen in school settings (Barrett & Webster, 2014).

References

Adorno, T. (1976). *Introduction to the sociology of music*. New York: Seabury Press.

Amabile, T. M. (1982). A consensual assessment technique. *Journal of Personality and Social Psychology, 43*, 997–1013.

Bamberger, J. (1977). In search of a tune. *The Arts and Cognition*, 284–319.

Bamberger, J. (2013). *Discovering the musical mind*. New York: Oxford University Press.

Barbot, B., Besançon, M., & Lubart, T. (2015). Creative potential in educational settings: Its nature, measure, and nurture. *Education 3–13, 43*(4), 371–381. http://doi.org/10.1080/03004279.2015.1020643.

Barbot, B., & Lubart, T. (2012). Creative thinking in music: Its nature and assessment through musical exploratory behaviors. *Psychology of Aesthetics, Creativity, and the Arts, 6*(3), 231.

Barbot, B., Lubart, T. I., & Besançon, M. (2016). "Peaks, slumps, and bumps": Individual differences in the development of creativity in children and adolescents. *New Directions for Child and Adolescent Development, 2016*(151), 33–45.

Barbot, B., Randi, J., Tan, M., Levenson, C., Friedlaender, L., & Grigorenko, E. L. (2013). From perception to creative writing: A multi-method pilot study of a visual literacy instructional approach. *Learning and Individual Differences, 28*, 167–176. https://doi.org/10.1016/j.lindif.2012.09.003.

Barbot, B., Tan, M., & Grigorenko, E. L. (2013). The genetics of creativity: The generative and receptive sides of the creativity equation. In O. Vartanian, A. Bristol, & J. C. Kaufman (Eds.), *The neuroscience of creativity* (pp. 71–93). Cambridge, MA: MIT Press.

Barbot, B., & Tinio, P. P. L. (2015). Where is the "g" in "creativity"? a speciali-zation-differentiation hypothesis. *Frontiers in Human Neuroscience, 8*(1041) https://doi.org/10.3389/fnhum.2014.01041.

Barrett, F. J. (2000). Cultivating an aesthetic of unfolding: Jazz improvisation as a self-organizing system. In S. Linstead & H. Hopfl (Eds.), *The aesthetics of organization* (pp. 228–245). London: Sage Publications.

Barrett, J. R., & Webster, P. R. (2014). *The musical experience: Rethinking music teaching and learning.* New York: Oxford University Press.

Barrett, M., & Tafuri, J. (2012). Creative meaning-making in infants' and young children's musical cultures. In G. McPherson & G. Welch (Eds.), *Oxford handbook of music education* (Vol. 1, pp. 296–313). New York: Oxford University Press.

Bashwiner, D. M., Wertz, C. J., Flores, R. A., & Jung, R. E. (2016). Musical creativity "revealed" in brain structure: Interplay between motor, default mode, and limbic networks. *Scientific Reports, 6,* 20482. https://doi.org/10.1038/srep20482.

Beaty, R. E. (2015). The neuroscience of musical improvisation. *Neuroscience and Biobehavioral Reviews, 51,* 108–117.

Bigand, E., & Poulin-Charronnat, B. (2006). Are we "experienced listeners"? A review of the musical capacities that do not depend on formal musical training. *Cognition, 100*(1), 100–130.

Brand, E. (2000). Children's mental musical organisations as highlighted by their singing errors. *Psychology of Music, 28*(1), 62–80.

Brinkman, D. J. (1999). Problem finding, creativity style and the musical compositions of high school students. *The Journal of Creative Behavior, 33*(1), 62–68.

Burnard, P. (1999). *Into different worlds (the soundings project): A study of children's experience of musical improvisation and composition.* Unpublished doctoral dissertation, University of Reading, UK.

Burnard, P. (2012). *Musical creativities in practice.* Oxford, UK: Oxford University Press.

Coleman, S. N. (1922). *Creative music for children.* New York: Putnam.

Csikszentmihalyi, M. (1988). Motivation and creativity: Toward a synthesis of structural and energistic approaches to cognition. *New Ideas in Psychology, 6*(2), 159–176.

Davidson, L., & Welsh, P. (1988). From collections to structure: The developmental path of tonal thinking. In J. Sloboda (Ed.), *Generative processes in music: The psychology of performance, improvisation and composition* (pp. 260–285). Oxford, UK: Oxford University Press.

Dunn, R. E. (1997). Creative thinking and music listening. *Research Studies in Music Education, 8*(1), 42–55.

Eisenberg, J., & Thompson, W. F. (2003). A matter of taste: Evaluating improvised music. *Creativity Research Journal, 15*(2–3), 287–296.

Feinberg, J. (1973). *Social philosophy.* Englewood Cliffs, NJ: Prentice-Hall.

Folkestad, G. (2012). Digital tools and discourse in music: The ecology of composition. In D. Hargreaves, D. Miell, & R. MacDonald (Eds.), *Musical imaginations: Multidisciplinary perspectives on creativity, performance and perception* (pp. 193–206). New York: Oxford University Press.

Folkestad, G., Hargreaves, D. J., & Lindström, B. (1998). Compositional strategies in computer-based music-making. *British Journal of Music Education, 15*(01), 83–97.

Gardner, H. (1983). *Frames of mind: The theory of multiple intelligences.* New York: Basic Books.

Getzels, J. W. (1964). Creative thinking, problem-solving, and instruction. In E. R. Hilgard (Ed.), *Theories of learning and instruction* (pp. 240–267). Chicago: University of Chicago Press.

Getzels, J. W., & Csikszentmihalyi, M. (1976). *The creative vision: A longitudinal study of problem finding in art.* New York: Wiley.

Glăveanu, V., Lubart, T., Bonnardel, N., Botella, M., Biaisi, P.-M. de, Desainte-Catherine, M., Zenasni, F. (2013). Creativity as action: Findings from five creative domains. *Frontiers in Psychology, 4.* http://doi.org/10.3389/fpsyg.2013.00176.

Guilford, J. P. (1950). Creativity. *The American Psychologist, 4,* 444–454. https://doi.org/10.1037/h0063487.

Guilford, J. P. (1967). *The nature of human intelligence.* New York: McGraw-Hill.

Hargreaves, D., Miell, D., & MacDonald, R. (2012). *Musical imaginations: Multidisciplinary perspectives on creativity, performance and perception.* Oxford: Oxford University Press.

Harlow, J., & Barbot, B. (2015, May 9). *MET: Compositional styles of novice musical composers and their effects on creative products.* Paper presented at the Pace University Psychology Conference, New York, NY.

Hewitt, A. (2009). Some features of children's composing in a computer-based environment: The influence of age, task familiarity and formal instrumental music instruction. *Journal of Music, Technology & Education, 2*(1), 5–24.

Hickey, M. (1997). Understanding children's musical creative thinking processes through the qualitative analysis of their MIDI data. *Bulletin of the Council for Research in Music Education, 131,* 29–30.

Kaschub, M., & Smith, J. (2009). *Minds on music: Composition for creative and critical thinking*. Lanham, MD: R&L Education.

Kaufman, J. C., & Baer, J. (2004). The Amusement Park Theoretical (APT) model of creativity. *Korean Journal of Thinking and Problem Solving, 14*, 15–25.

Kogan, N. (1994). On aesthetics and its origins: Some psychobiological and evolutionary considerations. *Social Research, 61*(1), 139–165.

Kratus, J. (1989). A time analysis of the compositional processes used by children ages 7 to 11. *Journal of Research in Music Education, 37*(1), 5–20.

Kratus, J. (1990). Structuring the music curriculum for creative learning. *Music Educators Journal, 76*(9), 33–37.

Kratus, J. (2012). Nurturing the song catchers: Philosophical issues in the teaching of music composition. In W. Bowman & A. Frega (Eds.), *Oxford handbook of philosophy of music education* (pp. 367–385). New York: Oxford University Press.

Leman, M. (1999). Music. In M. A. Runco & S. R. Pritzker (Eds.), *Encyclopedia of creativity* (pp. 285–296). San Diego: Elsevier.

Lubart, T. I., Zenasni, F., & Barbot, B. (2013). Creative potential and its measurement. *International Journal of Talent Development and Creativity, 1*(2), 41–51.

McPherson, G. E. (2000). Commitment and practice: Key ingredients for achievement during the early stages of learning a musical instrument. *Bulletin of the Council for Research in Music Education, 147*, 122–127.

Moorhead, G., & Pond, D. (1941–1944). *Music of young children*. Santa Barbara, CA: Pillsbury Foundation for Advancement of Music Education.

Nevels, D. L. (2013). Using music software in the compositional process: A case study of electronic music composition. *Journal of Music, Technology & Education, 5*(3), 257–271.

Sawyer, K. (2012). *Explaining creativity: The science of human innovation* (2nd ed.). New York: Oxford University Press.

Seddon, F. A., & O'Neill, S. A. (2003). Creative thinking processes in adolescent computer-based composition: An analysis of strategies adopted and the influence of instrumental music training. *Music Education Research, 5*(2), 125–137.

Sloboda, J. A. (1985). *The musical mind: The cognitive psychology of music*. Oxford, UK: Oxford University Press.

Sloboda, J. A. (1988). *Generative processes in music: The psychology of performance, improvisation, and composition*. Oxford, UK: Clarendon Press and Oxford University Press.

Sternberg, R. J. (1988). *The nature of creativity: Contemporary psychological perspectives*. Cambridge, UK: Cambridge University Press.

Teplov, B. M., & Deprun, J. (1966). *Psychologie des aptitudes musicales*. Paris: Presses universitaires de France.

Tobias, E. (2012). Let's play! Learning music through video games and virtual worlds. In G. McPherson & G. Welch (Eds.), *Oxford handbook of music education* (Vol. 2, pp. 531–548). New York: Oxford University Press.

Velia, E. J., Irvin, M. D., Berendt, S., & Ramirez, E. E. (1999). The effects of music on mood and perception of a visual stimulus. *Psi Chi Journal of Undergraduate Research, 4*(3), 101–105.

Wallas, G. (1926). *The art of thought*. New York: Harcourt.

Webster, P. (1987). Conceptual bases for creative thinking in music. In C. Peery (Ed.), *Music and child development* (pp. 158–174). New York: Springer-Verlag.

Webster, P. R. (1996). Creativity as creative thinking. In G. Spruce (Ed.), *Teaching music* (pp. 87–97). London: Routledge.

Webster, P. R. (2002). Creative thinking in music: Advancing a model. In T. Sullivan & L. Willingham, (Eds.), *Creativity and music education* (pp. 16–33). Edmonton, AB: Canadian Music Educators' Association.

Webster, P. R. (2011). Key research in music technology and music teaching and learning. *Journal of Music, Technology and Education, 4*(2/3), 115–130. https://doi.org/10.1386/jmte.4.2-3.115_1.

Webster, P. R. (2012). Toward a pedagogy of revision: Guiding middle school students' music compositions. In O. Odena (Ed.), *Musical creativity: Insights from music education research* (pp. 93–112). Farnham, Surrey, England: Ashgate.

Webster, P. R. (2016a). Pathways to the study of music composition by preschool to precollege students. In S. Hallam, I. Cross, & M. Thaut (Eds.), *The Oxford handbook of music psychology* (2nd ed., pp. 681–708). Oxford, UK: Oxford University Press.

Webster, P. R. (2016b). Creative thinking in music, twenty-five years on. *Music Educators Journal, 102*(3), 26–32.

Webster, P. R., & Hickey, M. (1995). Challenging children to think creatively. *General Music Today, 8*(3), 4–10.

Wiggins, J. (2015). *Teaching for musical understanding* (3rd ed.). New York: Oxford University Press.

11

The Creative Process in Lead Sheet Composition

Daniel Martín, François Pachet and Benjamin Frantz

Writing a Good Song

Songs invade our daily lives to such an extent that musical taste is now considered as a trait of our personality in Western societies (Rentfrow, 2012). Writing a good song is a highly delicate endeavor. Good songs achieve a subtle balance of melody, harmony, rhythm, lyrics, and sound, as well as many other factors such as the voice of the singer, the arrangements, orchestration, production, not to mention marketing and many other social factors.

Some studies claim that it is possible to predict the success of a song based on objective data (editorial, acoustic, social), but these claims are highly debatable (Pachet, 2011). In particular, social pressure has been shown to be a determining factor in explaining non-uniform distribution of taste in our societies (Salganik, Dodds, & Watts, 2006). As a consequence, and in spite of a recent burst in the study of the music

D. Martín (✉) · F. Pachet · B. Frantz
Paris, France

© The Author(s) 2018
T. Lubart (ed.), *The Creative Process*, Palgrave Studies in Creativity and Culture,
https://doi.org/10.1057/978-1-137-50563-7_11

275

composition process (Collins, 2012), very little is known about what are successful strategies for composing good songs.

Feedback in Song Writing

During the composition process of a song,[1] many types of interactions take place. First, when there are several composers writing a song (usually in duos like Lennon and McCartney), they work collaboratively to exchange ideas, or to perform trial-and-error explorations. Also, during rehearsals, performers report feedback to composers; either explicitly, by commenting certain parts of a song or suggesting changes, or implicitly, by performing the song differently from the original composers' instructions. Later, when the composition is performed live, the audience evaluates implicitly the composition, for example, by applauding effusively if they like it or less warmly if they do not. Eventually, the number of hits, or downloads from music sites, also gives some sort of feedback from the very end of the creation chain.

Intuitions

Several ethnographic studies have been conducted to study the composition process for contemporary music genres: studies on collaborative composition (Donin, 2016), as well as on the influence of performers in a composition (Ahren, 2015), but very little work has addressed popular music composition. The intuitions and the work described in this chapter originate mostly from the experience of the second author in song composition, in the jazz and pop genres. Many insights were collected during the song composition process taking place between 2011 and 2013, which led to the publication of two music albums, one in French pop (Pachet & Diran, 2014) and one in jazz (Pachet & d'Inverno, 2014). In order

---

[1]Although it is difficult to draw a line between experimental and popular music, we refer here to popular songs, i.e., songs that are composed with the aim of being performed publicly, distributed, sold, and more generally aiming at pleasing a specific audience.

Fig. 11.1 Interaction taking place during the song composition process, leading to the album Pachet and d'Inverno (2014). *Note* Mark Inverno (left) and François Pachet (right)

Fig. 11.2 Interaction taking place during the song composition process, leading to the album Pachet and Diran (2014). *Note* Catherine Diran (left), François Pachet (center) and Jean-Christophe Urbain (right)

to study the composition process at stake, composition involved 2 people in each case, and all the interactions were video taped (see, e.g., Figs. 11.1 and 11.2). This huge amount of recordings has not yet been fully analyzed, but has provided various insights about the song composition process.

Consensus without understanding. One of the most interesting observations is the fact that although there are usually many disagreements along the way (related to the different taste of the co-composers for instance), there is always a striking moment when the song is finally right, when it grooves, where everything (melody, harmony, sounds) fits together, and when *everyone agrees* without discussion, but also without knowing *why it works*. The same feeling can be observed by crosswordists guessing finally a tricky definition: they know they found it, they do not have to check the solution, but they do not know why, i.e., there is no clear validation process (in contrast to, e.g., a math problem, which usually contains a validation procedure).

Creating versus evaluating. One of the striking forces of co-composition is that the composition dialog enables the actors to switch between 2 mental modes: a creative mode and an evaluation mode. The creating mode is needed when there is a difficulty and a solution has to be found: one of the co-composers has to find a way out. The evaluation mode consists of revisiting the other co-composer's idea and correcting it, changing it or challenging it in some way. When composition is conducted alone, the composer has to switch constantly between these 2 modes, which is cognitively the most challenging aspect of music composition (and arguably of creation in general).

The importance of lead sheets. In jazz and in pop music, lead sheets play a major role. Of course, some pop song composers do not write explicitly lead sheets. For instance, Paul McCartney claimed always that he did not know how to write music. Same for Django Reinhardt, composer of the famous tune *Nuages*. However, even when they are implicit, lead sheets are the primary form of song creation: a melody with a chord grid. The composition process can be envisaged, at least for a first approximation as being lead sheet based.

However interesting, the analysis of such composition dialogues does not provide quantitative data about the strategies involved.

All interactions happening during the composition process can be defined as different forms of feedback. This is the motivation behind the experiment described here, in which we assess the impact of feedback on the quality of a music composition.

The concept of feedback, and more concretely, peer-feedback, which refers to feedback provided between equals, has become popular due to the increase of e-learning systems and online social networking sites (Dominic & Francis, 2013). In such sites, internet users exchange ideas about a given subject, e.g., music composition (Settles & Dow, 2013), or music production (Salavuo, 2006). In these sites, users collaborate to compose and produce music respectively. Also, online courses like MOOCS are becoming more and more popular (September, 2013).

Similarly, in the pedagogical domain, new teaching methods are emerging in which students receive feedback from their peers, rather than only from the teacher. Peer-feedback offers several benefits in education. Rollinson (2005) shows that peers can provide useful feedback, of a different nature (e.g., more informal) than the ones that a teacher can provide and that peer-feedback enforces collaboration between students and helps them to become more critical. Sadler and Good (2006) state that peer-grading leads to many benefits when students receive training by the teacher on grading. Similarly, Lam (2010) shows the benefits of training students for peer-reviewing in the context of an ESL (English as Second Language) class.

In order to evaluate the impact of peer-feedback on the *quality* of a song, we restricted our study to a specific type of feedback. In this experiment, feedback is provided in a way that is similar to *corrections* made in a learning context like ESL redacting: by proposing specific modifications in certain parts of the whole work and possibly commenting them.

Even though text correction and music composition reviewing are very different tasks, they are both a way of providing feedback. Therefore, in this experiment, feedback is suggested by peers in the form of modifications of certain parts of the composition.

Feedback in this experiment is anonymous, because we analyze the quality of the musical suggestions regardless of the relationship between composers and commentators, as it has been shown that relationships between peers can influence collaborative music composition (Miell & MacDonald, 2000).

Lead Sheets

This chapter focuses on lead sheets. A lead sheet consists of a melody, most of the time monophonic as it is usually intended to be sung, and a sequence of chord symbols representing the harmony (e.g., Amaj7, Dm7, E7b9, etc.). Lead sheets are then arranged, orchestrated, and more generally realized and produced. Such a viewpoint is, of course reductionist, as it is probable that dimensions that are not represented in lead sheets, such as sound, instrument or voice timbre, do influence the composition process (albeit in still unknown ways). However, we restrict our experiment to lead sheets, as they are a primary form of pop song composition, and have an acknowledged existence.[2]

Lead sheets are widely used in pop music, as well as in jazz, bossanova, and many other popular music genres. Figure 11.3 shows the lead sheet of *Pretty Late*, a composition from Pachet and d'Inverno (2014). Jazz standards are usually played in jam sessions, where musicians can play together with other musicians without knowing each other. Musicians use the lead sheet as a guide. A typical jazz ensemble is composed by a musician playing the melody (e.g., a saxophone or a trumpet), another (usually a guitarist or pianist) playing the harmony defined by the chord grid, a bassist, also following the chord grid, and a drummer. Lead sheets are suitable for a music genre like jazz, in which musicians can play in a very free way. Even though the lead sheet represents only the essential information of a song, jazz musicians have enough knowledge to play the song by following it, even if they do not know the song. For example, a guitar player infers what chords to play from the lead sheet's chord grid. Then, he decides other dimensions of performance, such as which notes to play, where in the fingerboard, the tempo, etc.

---

[2]Lead sheets, for instance, are the primary assets of music publishing companies, which is a tangible sign that they somehow represent the essence of a song, regardless of its possible interpretations.

Fig. 11.3   Pretty Late, a composition from Pachet and d'Inverno (2014)

The experiment we describe below is based on an online lead sheet editor (Martín, Neullas, & Pachet, 2015) that was used in particular for populating a large lead sheet database (Pachet, Suzda, & Martín, 2013).

Experiment on Feedback in Lead Sheet Composition

To what extent can peer-feedback affect the quality of a music composition? How does musical experience influence the quality of feedback during the composition process? To answer these questions, we propose an experiment in which participants compose songs, provide feedback to other participants and try to improve their own initial composition.

Participants are assigned randomly into two groups: the control group (G1) does not receive feedback, so participants from this group have no external help when improving the composition. Participants from the experimental group (G2) receive feedback from two other participants and can use it to improve their composition.

To measure the quality of a music composition, we take into account two types of quality: (1) *subjective quality* is provided by the composer of the song; (2) *consensual quality* is obtained by social consensus, i.e., by aggregating the opinions of several participants. Furthermore, we estimate the composition experience level of each participant, as well as a more general musical level, by asking him or her to complete a questionnaire before starting the experiment.

Experiment Protocol

The experiment is performed online, and participants are recruited from mailing lists of composition in jazz and pop music. Consider now each phase of the study.

Experience questionnaire. When participants are logged in the online tool, they are asked to complete a questionnaire about their music skills in performing and composing. For example, they are asked how many years they have studied music theory, how many years they have been playing in a band, which style of music they like more, or how often they compose (see Fig. 11.4).

Song composition. In a first phase, participants have to compose an 8-bar lead sheet using the online editor. Participants cannot add or delete bars. However, they are free to choose the time signature, the tonality, and the tempo. Participants have to enter the melody and the chord grid with chord symbols (e.g., Cmaj7, Dm, etc.). Participants can listen to their composition with a basic MIDI player. Once they are done, they cannot edit the song anymore. Next, they answer a questionnaire about their confidence in the quality, complexity, and satisfaction concerning their composition.

Providing feedback. In this phase, each participant is assigned randomly to another participant's composition, and is asked to make suggestions to improve it. These suggestions are expressed as modifications

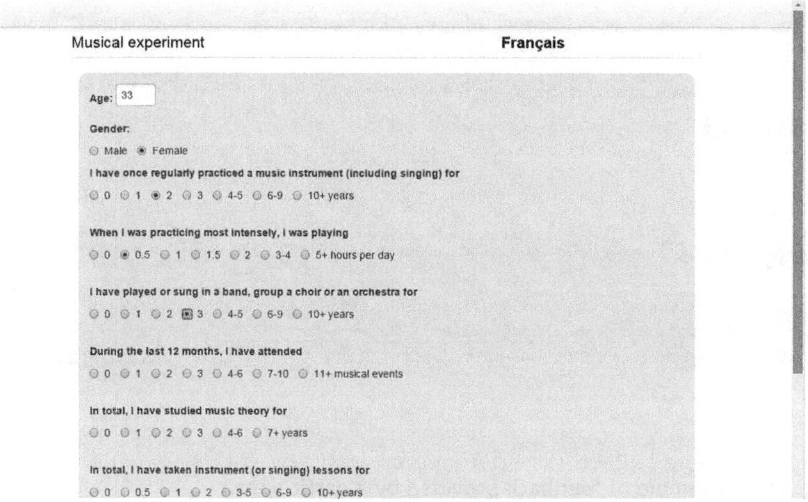

Fig. 11.4   Questionnaire completed by participants

of notes or chord symbols on a specific region of the composition with a duration of one or two bars. As many suggestions as wanted can be issued as long as they do not overlap with each other. To make a suggestion, participants must choose the bar(s) to modify, then they can change the notes and the chord symbols. Optionally, they can write a text comment explaining their changes (see Fig. 11.5).

Song improvement. In this phase of the experiment, participants are asked to improve their initial song. Those from the control group (G1) try to improve it by themselves with no help from peers, whereas those from the experimental group (G2) are invited to review the suggestions from other participants, play them, and accept or reject them. They can also modify freely their song. At this point, each participant has produced two versions the song: the original and the final one. Figure 11.6 shows an example of an original composition from the experiment and Fig. 11.7 the same composition after improvement. Once they are finished, they answer a questionnaire about their confidence in their own improvement and their opinion on the suggestions received.

Evaluation. In the final phase of the experiment, participants must evaluate at least 5 pairs of songs of other participants by listening to

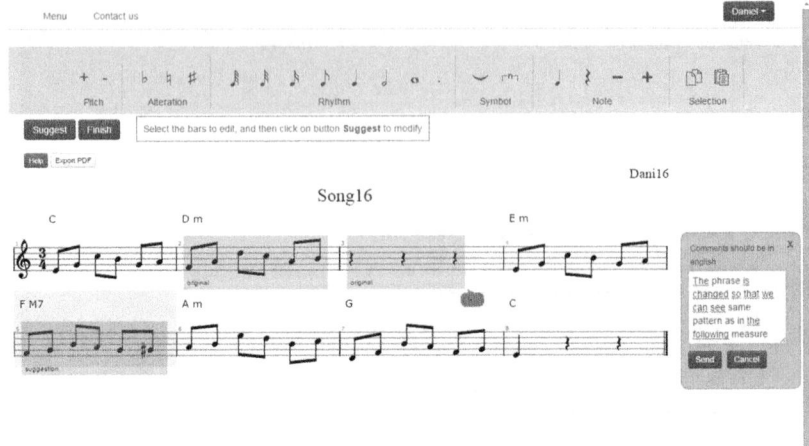

Fig. 11.5 Example of feedback provided by a participant

Fig. 11.6 Example of an original composition from a participant

them and giving them a note between 0 and 100 according to how much they like it. Participants do not know which is the original and the improved song. One of the versions is presented as *song A* and the other as *song B* and this assignment is performed randomly. Each pair of songs is presented separately, through a pagination system designed so that participants are forced to listen and evaluate songs in a short amount of time. We want to avoid having participants listen to a song

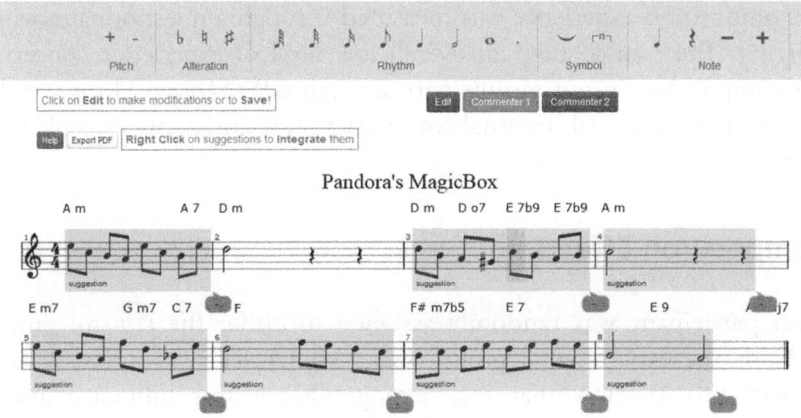

Fig. 11.7 Example of an improved composition

once and then going back to it another day to listen to it again and evaluate it, because this could bias the results, as previous research has shown that previous exposure to a melody has an effect on preference (Peretz, Gaudreau, & Bonnel, 1998).

Results

In this section we describe the results obtained from each phase of the experiment.

Population

Sixty-six participants completed the experiment (68% men and 32% women). Mean age was 29.2 years, ranging from 19 to 61. Musical experience was measured through a questionnaire with 7 items. The scale has a satisfactory sensitivity with an observed range from 7 to 41 (out of 0–42) and we observed a mean of 28.7 with a standard deviation (SD) of 8.9. Internal consistency was satisfactory (Cronbach's alpha = .82).

Composition experience was measured through a questionnaire with 5 items. The results show an overall low level of experience concerning composition in our sample with a mean 6.9 ($SD = 6.1$) on a scale ranging from 0 to 30). Internal consistency was satisfactory (Cronbach's alpha = .85).

Composition Effects

Each participant was randomly assigned to either the control group (G1) or the experimental group (G2). Müllensiefen, Gingras, Musil, Stewart (2014) show that sociodemographic factors influence musical skills. However, no significant differences were observed between the two groups in relation to age and gender, musical experience or composition experience.

Composition evaluations. During the evaluation step, we checked if participants had actually listened to the songs before evaluating them. Of the 1195 evaluations made, 219 were made without listening to the song. We removed those evaluations.

The songs were evaluated by an average of 8.8 different judges. The mean score of the evaluations made during the evaluation phase is 53.25 ($SD = 13.26$) on a scale ranging from 0 to 100. However, some judges might be stricter than others, and some songs might have been evaluated by a particularly strict or generous participant. To take into account the disparity in the judge ranking schemes, we standardized the evaluations to get z-scores where the mean and standard deviation used are based on all the evaluations made by a given participant. As a result, the mean of the standard scores is approximately equal to zero, and the standard deviation is approximately .50. It should be noted that this final score correlates strongly with the raw score ($r = .84$). This result indicates that we had enough evaluations for each song to avoid bias due to the disparity of the judges.

Original Composition. The questionnaire that participants were asked to complete after finishing the original composition included self-estimation questions about the quality, complexity, and satisfaction for their composition, with scales ranging from very bad/

simple/unsatisfied (0) to very good/complex/satisfied (6). Also, we asked participants to evaluate the time they spent to make their composition and if they used a musical instrument to help them to compose (and which instrument if they did).

Results show a mean quality of 2.8 ($SD = 1.5$), a mean complexity of 1.9 ($SD = 1.6$) and a mean satisfaction of 3.2 ($SD = 1.6$). Only the complexity is significantly different from the center of the scales, which is 3 ($T(65) = -5.27$; $p < .0001$). This means that the participants tend to judge their work as rather simple (low complexity). We observed also positive and significant correlations among these three measures, ranging from $r = .41$ to $r = .80$.

During the suggestion phase, we asked the participants to rate also the quality and complexity of the songs they had to comment. Each composition from the experimental group (G2) was commented by two participants. In the end, we obtained the score from the author and two other scores from two different commentators. Interestingly, there was no correlation between the scores from the original composer and the ones from the commentators ($r < .10$), but the two commentators did agree together on the quality ($r = .80$) and on the complexity ($r = .70$).

Moreover, from the judgments made during the evaluation phase (in which participants evaluate pairs of songs from other participants), the measurement of the consensual quality of each original song (standardized to z-scores) allows us to estimate the composition skill level of its author. Surprisingly, we observed that the quality of the original song is only marginally related to composition experience ($r = .18$, $p = .15$) or to musical experience ($r = .19$, $p = .12$). We asked participants if they used an instrument to help them in their composition. Results show a marginally significant effect in favor of the use of an instrument on the mean quality score ($T(64) = -.87$, $p = .38$).

The mean duration of the time taken to compose the song as evaluated by the participants was 30 minutes ($SD = 32$ min) ranging from 1 to 240 minutes. This evaluation is largely underestimated by the participants: the real duration calculated from the time spent on the composition software was significantly longer ($m = 67$ min; $T(65) = 4.20$, $p < .001$). The correlation between these two durations is not very high, but significant ($r = .46$, $p < .001$), indicating that the error of duration estimation is not the same for everyone. Interestingly, we observed that the quality of the original songs (from the evaluation phase) is not related to the time spent to compose, whether it is subjective ($r = .04$) or consensual quality ratings ($r = .03$). This result suggests that in a situation where there is no time constraint, the amount of time devoted to compose has no effect on its quality.

Finally, there was a difference in the consensual quality of the original song, obtained from the evaluation of several participants (.07 in G1 vs. $-.15$ in G2). This could, however, be explained by differences in the group of judges evaluating each song.

Suggestions. In the questionnaire completed after making the suggestions, participants were asked how much they think the song they were revising will be improved due to their modifications (on a 7-point Likert scale ranging from 0 "very little," to 6 "very much").

The participants from G2, the experimental group ($N = 30$), received two suggestions for their final composition. Once they finished, we asked them if the suggestions received were interesting (on a 7-point Likert scale ranging from 0 "very little," to 6 "very much"). Additionally, we recorded the number of suggestions they received, the number of suggestions they used, and the number of texts comments received. We ran a series of correlations between these measures and the improvement effect (the difference between the original song and the final song on the quality judgment score). None were significant, suggesting that neither the number of suggestions received nor the number of explanations for those suggestions is related systematically to the improvement of a song.

The suggestions received could concern the notes or the chords, and the receiver of the suggestions could choose to use them or not. Based on the 154 suggestions that were made, 38 concerned both notes and chords, 57 only notes, 44 only chords, and 15 were only text comments.

Final Composition. Overall, we can see that the control group, G1, does not improve significantly between the original song ($m = .07$) and the final song ($m = .12$) (improvement effect $= .05$, $T(35) = .94$, $p = .35$). However, we see a significant improvement for the experimental group, G2, between the original song ($m = -.15$) and the final song ($m = .08$) (improvement effect $= .23$, $T(29) = 2.47$, $p = .02$). See Fig. 11.8.

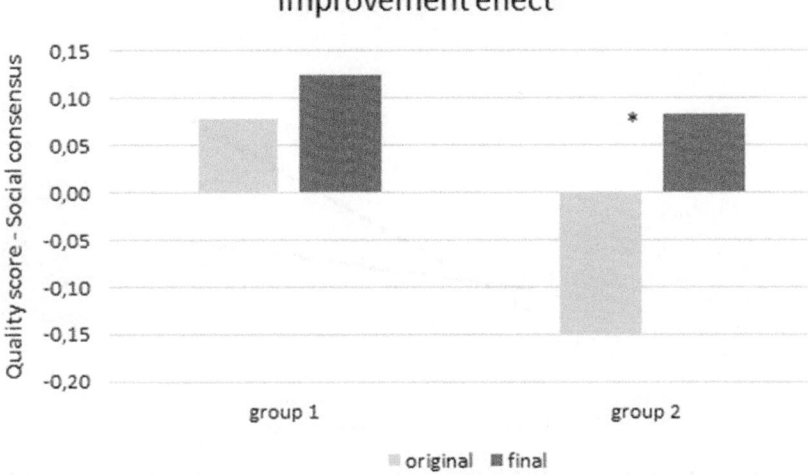

Fig. 11.8 Difference between the original song and the final song on the quality judgment score for the group without feedback (G1) and the group with feedback (G2)

We examined also the subjective evaluation of the participants concerning the improvement of their song. We constructed two composite scores. One from the self-evaluation scales of the original song (quality, complexity and satisfaction) and one from the self-evaluation scales of the final song (quality, complexity and satisfaction). The internal consistency of those composite scores is satisfactory (the two Cronbach' alphas are above .81). Then, we conducted a mixed *between participants* (control and experimental groups) × *within participants* (original and final song) analysis of variance. We observed a significant interaction between groups and songs ($F(1, 64) = 7.07$, $p = .01$). To explore this interaction, we used a post hoc analysis with Tukey HSD tests. Results show that participants who received suggestions had a significant improvement between the original and final song ($p < .001$), whereas the control group had no improvement ($p = .49$) (see Fig. 11.9).

When evaluating songs, users did not know which song was the original and which one was the final, as the order of the songs was determined randomly. This was a design decision to avoid the fact that

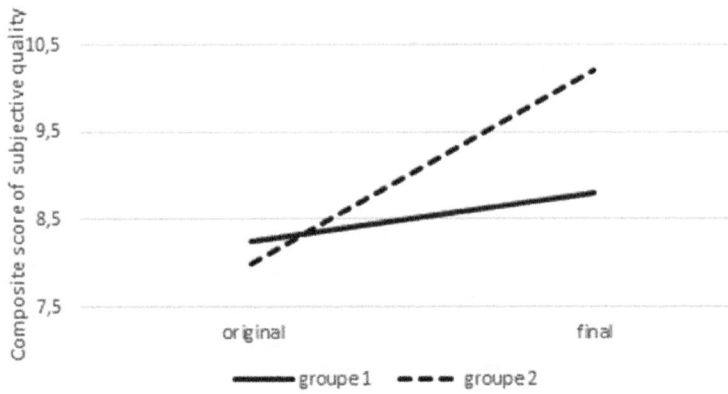

Fig. 11.9 Self-rated quality of the original and final songs for the group without feedback (G1) and the group with feedback (G2)

participants could tend to rate better the final song, as it is supposed to be improved. Additionally, we wanted to ensure that songs were not better rated just because they had more modifications. To check this point, we used a *melodic similarity algorithm* from Urbano, Lloréns, Morato, and Sánchez-Cuadrado (2011) to estimate the similarity between original and final songs. The correlations between the amount of similarity and the improvement effect based both on the composer's subjective opinion and on the scores from the judges are low ($r = -.36$, $p = .003$ and $r = -.19$, $p = .13$), which suggests that the improvement is not related to the dissimilarity between the two versions.

Also, we see no relation between the subjective improvement and the number of suggestions received ($r = .07$). However, a significant correlation appears when we look at the number of suggestions used ($r = .47$). This result suggests that the subjects proceeded to a selection of the suggestions provided and did not simply integrate them. Due to the limited size of the sample, we took extra care when looking at the scatter plot to ensure the absence of atypical subjects.

Based on the 154 suggestions received by the subjects of group 2, a multiple linear regression was calculated to predict the subjective improvement effect based on (a) the subjective quality of the original song, (b) whether the suggestion concerns notes, (c) whether the suggestion concerns chords, and (d) the number of suggestions used. A significant regression equation was found ($F(4, 149) = 29.21$, $p < .001$) with an $R^2 = .44$). As expected, the subjective quality of the original song is the strongest predictor ($\beta = .57$, $p < .001$). Also as expected, the number of integrated suggestions is a significant predictor ($\beta = .17$, $p < .001$).

Interesting results were obtained concerning the type of suggestions. When suggestions concern chords, they are a significant predictor, with the same weight as the number of integrated suggestions ($\beta = .19$, $p < .001$), but it is not the case when the suggestions concern notes ($\beta = -.01$, $p = .92$). This is interesting because it suggests that improvement of songs is made through chord modifications rather than melodic changes. Confirming our previous results, adding the composition experience level or music experience level of the commentator to the equation appears to have no effect on the improvement and does not change the equation ($\beta = .07$, $p = .38$ and $\beta = .11$, $p = .14$, respectively).

Lead sheet editor. The software used was developed specifically for the experiment and we asked participants whether it was frustrating (0) or helpful (6) to compose with it. Results show a mean of 3.13 after the first composition and 3.41 after the final composition (the difference is not significant), which means that even if the online editor was not especially helpful, it did not hinder the composition process.

Experience effect on evaluations. We assumed that the consensual quality is a reliable evaluation. Similarly, Hickey (2001) relied on *consensual assessment* to measure the creativity of children's musical compositions. Nevertheless, he found that composers evaluated differently from other groups such as music teachers, music theorists, etc. In our case, we examined whether musical experience has an impact on the way participants judge songs from other participants, and we divided our sample of participants into two groups according to their experience as a musician (based on the median). We divided also our sample of songs according to the musical-experience level of the authors. A two-way ANOVA was then used to explore the effect of the experience of the judges according to the experience of the composer. Results show an interaction between these two variables ($F(1, 61) = 7.63$, $p = .007$)

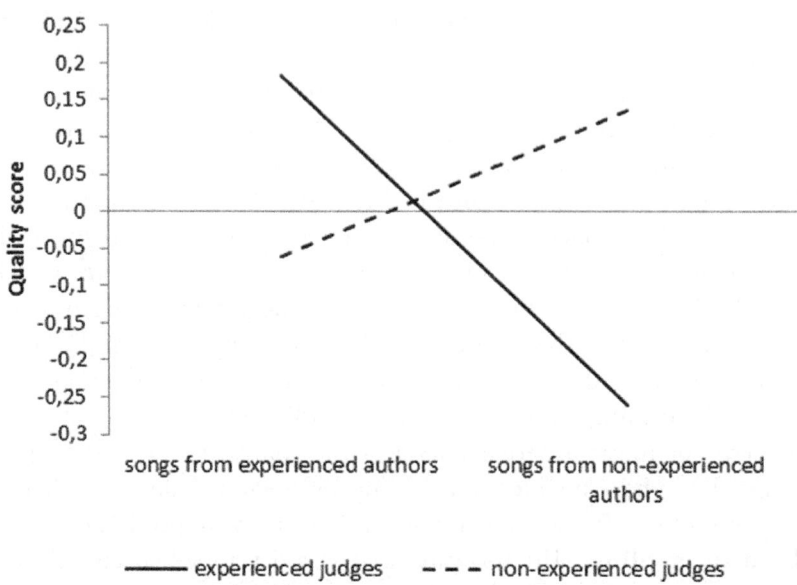

Fig. 11.10 Interaction between the experience of the author and the experience of the judges on the quality score

as illustrated in Fig. 11.10; experienced judges give high scores to songs from experienced authors and low scores to songs from non-experienced authors. It is exactly the opposite for the non-experienced judges. This means that participants tend to prefer compositions from other participants with similar experience, which could explain partly the difference in the evaluation of the original songs in G1 and G2.

Conclusions

The aim of this experiment was primarily to examine quantitatively the impact of peer feedback in music composition and second to assess the influence of participants' experience level as musicians or composers on the whole process. Before any improvement or suggestions, participants had to write their first song. Interestingly, results show that participants' previous experience in composition did not impact the quality of their songs. The same pattern was also found concerning the participants' previous experience as a musician. These two results suggest that the quality of a song (based on social consensus) does not really tap musicality but something else, presumably creativity (Frese, Teng, & Wijnen, 1999).

Results show that composers who received feedback (G2) evaluated more positively the improved song compared to the original one, meaning that they were satisfied with the improvement they made. Furthermore, the evaluation based on social consensus led to greater improvement for G2. Hence, participants who received feedback not only felt that they had composed a better song after the improvement step, but they actually did. This finding suggests that improvements in a music composition may be achieved even without real collaboration involving dialogue and active interaction, but by simple suggestions on a single occasion.

Because there was a difference on the evaluation of the original songs between G1 and G2, we wanted to verify whether musical experience can make a difference when evaluating songs and we found out that participants tend to prefer songs composed by other participants with similar musical experience.

Future work may determine better the influence of participants' musical experience, in particular on consensually assessed quality. This could be done by checking when songs are most improved in the composition process, taking into account the experience of composers, commentators, and judges. Furthermore, we could check if suggestions from experienced commentators are more likely to be used from inexperienced composers, or whether experienced composers accept usually suggestions of other composers, and how this may affect the improvement of the song.

Studies as described in Bigand and Poulin-Charronnat (2006) suggest that some musical skills may be acquired by mere exposure to music, with no need for training. So, in addition to looking at evaluators' musical experience and composition experience, we could check their *listening experience* as well.

A final question that can be addressed concerns the impact of peer-feedback on creativity (as opposed to quality). Some studies have attempted to measure creative thinking in music, such as Barbot and Lubart (2012) (see Chapter 10 in this volume). Other studies, conducted in the field of design, show that previous exposure to ideas has an effect on the design creativity. Smith, Ward, and Schumacher (1993) show that exposure to previous examples can have negative effects such as constraining the generation of ideas. Bonnardel and Marmèche (2005) point out that experienced designers use exposed examples better than novices in their creations (see Chapter 9 in this volume). Following these investigations, we could evaluate how exposure to musical suggestions influences composers' creativity, notably the originality of compositions.

Acknowledgements This research was conducted within the Flow-Machines project funded by the ERC under the European Union's 7th Framework Programme (FP/2007–2013)/ERC Grant Agreement n. 291156, and by the Praise project (EU FP7 number 388770), a collaborative project funded by the European Commission under program FP7-ICT-2011-8.

References

Ahren, T. (2015, 10). *Versions, variants, and the performatives of the score: Traces of performances in the texts of Anton Webern's music*. Retrieved from http://medias.ircam.fr/xc1449e.

Barbot, B., & Lubart, T. (2012). Creative thinking in music: Its nature and assessment through musical exploratory behaviors. *Psychology of Aesthetics, Creativity, and the Arts, 6*(3), 231.

Bigand, E., & Poulin-Charronnat, B. (2006). Are we "experienced listeners"? A review of the musical capacities that do not depend on formal musical training. *Cognition, 100*(1), 100–130.

Bonnardel, N., & Marmèche, E. (2005). Towards supporting evocation processes in creative design: A cognitive approach. *International Journal of Human-Computer Studies, 63*(4), 422–435.

Collins, D. (2012). *The act of musical composition: Studies in the creative process.* Farnham: Ashgate.

Dominic, M., & Francis, S. (2013). An assessment of popular e-learning systems via Felder-Silverman model and a comprehensive-learning system using the tools on web 2.0. *International Journal of Modern Education and Computer Science, 5*(11), 1.

Donin, N. (2016). Domesticating gesture: The collaborative creative process of Florence Baschet's StreicherKreis for 'augmented' string quartet (2006–2008). In E. Clarke & M. Doffman (Eds.), *Creativity, improvisation and collaboration: Perspectives on the performance of contemporary music*. New York: Oxford University Press.

Frese, M., Teng, E., & Wijnen, C. J. (1999). Helping to improve suggestion systems: Predictors of making suggestions in companies. *Journal of Organizational Behavior, 20*(7), 1139–1155.

Hickey, M. (2001). An application of Amabile's consensual assessment technique for rating the creativity of children's musical compositions. *Journal of Research in Music Education, 49*(3), 234–244.

Lam, R. (2010). A peer review training workshop: Coaching students to give and evaluate peer feedback. *TESL Canada Journal, 27*(2), 114.

Martín, D., Neullas, T., & Pachet, F. (2015). Leadsheetjs: A javascript library for online lead sheet editing. In *First International Conference on Technologies for Music Notation and Representation (Tenor)*. Paris, France.

Miell, D., & MacDonald, R. (2000). Children's creative collaborations: The importance of friendship when working together on a musical composition. *Social Development, 9*(3), 348–369.

Müllensiefen, D., Gingras, B., Musil, J., & Stewart, L. (2014). The musicality of non-musicians: An index for assessing musical sophistication in the general population. *PLoS One, 9*(2), e89642.

Pachet, F. (2011). Hit song science. In Tao, Tzanetakis, & Ogihara (Eds.), *Music data mining* (pp. 305–326). Boca Raton, FL: Chapman & Hall/CRC Press.

Pachet, F., & Diran, C. (2014). *Marie-claire*. http://www.cdbaby.com/AlbumDetails.aspx?AlbumID=marieclaire.

Pachet, F., & d'Inverno, M. (2014). *Count on it*. http://www.cdbaby.com/cd/markdinvernoquintet?SourceCode=widgetbaby.

Pachet, F., Suzda, J., & Martín, D. (2013, November). A comprehensive online database of machine-readable lead sheets for jazz standards. In *ISMIR* (pp. 275–280). Curitiba (Brazil).

Peretz, I., Gaudreau, D., & Bonnel, A.-M. (1998). Exposure effects on music preference and recognition. *Memory & Cognition, 26*(5), 884–902.

Rentfrow, P. J. (2012). The role of music in everyday life: Current directions in the social psychology of music. *Social and Personality Psychology Compass, 6*(5), 402–416.

Rollinson, P. (2005). Using peer feedback in the ESL writing class. *ELT Journal, 59*(1), 23–30.

Sadler, P. M., & Good, E. (2006). The impact of self-and peer-grading on student learning. *Educational Assessment, 11*(1), 1–31.

Salavuo, M. (2006, 11). Open and informal online communities as forums of collaborative musical activities and learning. *British Journal of Music Education, 23*, 253–271. https://doi.org/10.1017/s0265051706007042.

Salganik, M. J., Dodds, P. S., & Watts, D. J. (2006). Experimental study of inequality and unpredictability in an artificial cultural market. *Science, 311*(5762), 854–856.

September, O. (2013). Behind the scenes with MOOCs: Berklee college of music's experience developing, running, and evaluating. *Continuing Higher Education Review, 77*, 137.

Settles, B., & Dow, S. (2013). Let's get together: The formation and success of online creative collaborations. In *Proceedings of the Sigchi Conference on Human Factors in Computing Systems* (pp. 2009–2018).

Smith, S. M., Ward, T. B., & Schumacher, J. S. (1993). Constraining effects of examples in a creative generation task. *Memory & Cognition, 21*(6), 837–845.

Urbano, J., Lloréns, J., Morato, J., & Sánchez-Cuadrado, S. (2011). Melodic similarity through shape similarity. In *Exploring music contents* (pp. 338–355). Berlin: Springer.

12

The Cultural Basis of the Creative Process: A Dual-Movement Framework

Vlad Petre Glăveanu

The question of the creative process has been at the heart of creativity studies for a number of decades. Trying to answer the 'how' of creativity, this question has been approached until now from a variety of psychological, neuro-psychological, and social science perspectives. However, despite numerous investigations—and the present volume testifies to this continued interest—there seems to be a lot to learn still when it comes to the creative process. This rather frustrating state of affairs takes us back to the very paradox of creativity research and its aim of putting 'order' and 'predictability' in an area that traditionally deals with the 'emergent' and 'unexpected'. The creative process results in new, original, and surprising outcomes and yet there is an expectation, at least within science, that we should uncover the exact steps leading to such outcomes. In this chapter, I will develop an optimistic view of our capacity to study creative processes and will do so

V. P. Glăveanu (✉)
Webster University Geneva, Bellevue, Switzerland
e-mail: glaveanu@webster.ch

© The Author(s) 2018
T. Lubart (ed.), *The Creative Process*, Palgrave Studies in Creativity and Culture,
https://doi.org/10.1057/978-1-137-50563-7_12

from a particular perspective—that of culture or, more specifically, of sociocultural psychology.

This perspective is paradoxical inasmuch as it places itself within psychology while advocating for an expanded notion of what is 'psychological'. To take the concrete example of creativity, the study of creative processes seems to have moved from examining (macro) stages of creating to understanding (micro) processes (Botella, Nelson, & Zenasni, 2016; Lubart, 2000). These micro-processes are largely assumed to be psychological in the sense that they take place in the mind and, more recently, are studied in the brain as well. They are thus both individualised and reduced to cognition first and foremost. To create means to combine, associate, diverge, converge, make analogies, use metaphors, so on and so forth. To create means thus to think, hence the relentless focus of creativity scholars on 'creative thinking'. What about creative doing?

We can say that creative thinking is a form of action or, at least, that it is incorporated within human creative action. But to reduce action to thinking alone ends up excluding everything that is beyond the individual mind and yet fully participates in creative processes, for example, social relations, objects, institutions, the body and, last but not least, culture. More importantly, it risks misconstruing mind as 'inside the head' and culture as 'outside the person'. In reality, mind and culture are *interdependent* (Shweder, 1990) and creative action is an excellent example of this interdependence.

The three premises I start from in this chapter are as follows:

1. The creative process is primarily a form of action and not (only) thinking;
2. Creative action involves both mind and culture in their interdependence;
3. The creative mind and its processes are, at once, psychological, material, and social—in other words, the human creative mind is *already cultural.*

In making the claims above I am trying to avoid longstanding and unproductive dichotomies that have marked the field of creativity studies since its inception (for details see Glăveanu, 2013), for instance, mind versus culture, thinking versus doing, idea generation versus idea implementation, self (creator) versus other (audience). Most of all, I am trying to avoid a definition of culture that considers it external to the person, embedded exclusively in social interactions, material and institutional arrangements, norms and values. In my—sociocultural and psychological—understanding, culture exists 'inside' *and* 'outside' the person and both the content and processes of our psychological life are impregnated by culture. Creativity makes no exception. In fact we can easily notice how relevant culture is for studying the creative process simply by looking at the approach taken within this book which is domain based. Domains of creative expression, such as the arts, sciences, design, etc., are *culturally constituted* as they involve specific material, social, and symbolic elements (think, for example, about the inter-related triad of person, field and domain in Csikszentmihalyi's (1998), systemic model). There is cultural specificity in how we express ourselves creatively not only across countries but within professional domains (Glăveanu & Lubart, 2018). In this sense, culture cannot be equated with national culture, as is often the case in cross-cultural psychology, but describes the wider socio-material-symbolic environment in which we grow, live, and create.

My aim in this chapter is twofold. On the one hand, I want to substantiate the cultural approach to creative processes. On the other, I want to review the contributions to this volume in light of this approach. What I hope to accomplish is a deeper understanding of the cultural basis and nature of creativity, one that doesn't exclude but integrates existing work and widens its scope. In doing so, I will start by selectively analysing what is 'cultural' (i.e., material, social, and symbolic) in the creative processes discussed in the previous chapters and how we can move from talking about creativity in intra-psychological terms to envisioning an 'extended' creative mind. Second, I will use the model of distributed creativity (Glăveanu, 2014) to examine forms of distribution involved in creativity in different domains, with particular focus on the situated nature/domain specificity of creative action. Finally, I will formulate a dual-movement cultural framework of the creative process that articulates the cyclical movement between immersion and detachment in creative work across different domains. In the end, I return to key questions about creativity, mind, and culture.

The 'Extended' Creative Mind

Whereas mental processes are crucial for the dynamic of creativity, it is equally true that the mind by itself cannot be called creative. For something to be creative it requires both some form of expression or externalisation (usually in oral, written or material form) and judgement or evaluation (by self and/or others, using cultural symbols and norms). At the extreme, we could think of ideas as potentially creative, but for creativity to occur ideas need to be translated into actions, and human action is, at all times, inter-action, as it mobilises social and material resources to accomplish its goals. In this sense, the creative mind can only be an 'extended' mind, meaning a mind that expands towards its environment. It is a mind whose borders are less well defined because it continues, through action, into the world. For example, many of the functions we traditionally consider 'psychological', such as memory or imagination, are in fact an accomplishment of the person–environment system. How do we remember and imagine? With the help of technology and objects, as well as with the help of other people. Isn't the physical or (increasingly) the electronic agenda for the day an 'external' memory device? These kinds of questions became central, in recent decades, for what is called distributed cognition (Hutchins, 1995) or the extended mind (Clark & Chalmers, 1998).

We can equally ask how creative processes can be (purely) intra-psychological. Is there any creative thinking happening in a vacuum, outside the material or social world? As many of the contributors to this volume noted about creativity in different domains, this is not the case. Toulouse (this volume) interestingly proposes the metaphor of a tree to discuss the creative process in art, a very 'organic' image of how creative work develops over time. Martín, Pachet and Frantz (this volume) offer a very social account of what it means to be a creative song composer. In this type of work, several interactions are crucial: with fellow composers, with performers, and with the audience. Feedback from others is, in their chapter, an essential part of achieving creative success. Piirto (this volume) makes a similar case for writers. Challenging the myth of the lone writer, she reviews evidence of how important group trust is for the process of writing. Friendships, competitions and critiques mark the exchanges between writers and shape their creative careers. This is even more strongly the case for screenplay writers. As Bourgeois-Bougrine and Glăveanu note in their chapter (this volume), scripts are essentially collaborative projects throughout the process of writing and this is so even when writers don't work together on the same script. As one of the participants states, a film has many midwives instead of being the 'child' of the director alone—from writers to producers, technicians, and the cast itself. Finally, Nelson (this volume) builds a similar argument for design, this time emphasising not only the social but the material/technological aspect. From co-design to distributed design and from boundary to intermediary objects, the creative process in design is clearly not only psychological but social and material. The question is, are sociomaterial aspects part of the process itself?

This is a difficult question to answer because it depends on the epistemological position we adopt. Upholding a strict Cartesian split between mind and matter would lead us to discuss social and material factors simply in terms of cognitive inputs processed and responded to by the person. In this sense, the sociomaterial aspect is present but just as an external variable, a source of stimulation for individual brains, and nothing else. On the contrary, from a sociocultural perspective, the social and material aspects are part of culture and culture itself connects mind and its environment. Discussions with others, the manipulation of objects, the use of technology, etc., are integral to an extended creative mind. They don't count only as a form of stimulation or as a prop, they don't just support or shape but come to *constitute* psychological processes. This is not meant to imply a form of social or material determinism that denies the mind and, by extension, the individual as an agent. On the contrary, the relation between person and environment is bidirectional and we are all both acquiring and contributing to culture through on-going processes of internalisation and externalisation (Moran & John-Steiner, 2003). According to this perspective, it would thus make more sense to talk about psycho-socio-material processes of creating (Glăveanu, 2011) and, instead of asking what is psychological, what is material, and what is social in creative work, focus on how these facets collaborate within creative action.

As follows, I will illustrate how a sociocultural approach to 'intra-psychological' creative processes looks like using some key processes uncovered in different chapters within this volume. Of special interest are: analogical thinking and associations/combinations, perspective-taking and flexibility, hypothesis formulation and imagination, and openness to experience. These are all arguably mental processes, some of them considered cognitive functions (e.g., analogical thinking), others labelled as personality traits (e.g., openness to experience). Although cognition and personality play undoubtedly a great part in creativity, as argued by many authors (see, for instance, Sternberg & Lubart, 1995), their functioning cannot be discussed separately from culture and its symbolic, social, and material dimensions. How can we achieve an integrative view?

Starting with the first example, that of *analogical thinking*, this process features primarily in Bonnardel, Wojtczuk, Gilles and Mazon's chapter on design (this volume), and builds on the A-CM model of Bonnardel (2000). In essence what this function points to is the capacity of designers, and creators more generally, to establish connections between their domain and other domains and to mentally transfer features from one domain to another. This is basically the logic of association, a very old law of psychology that offers the foundation for many theoretical models of creativity. It basically postulates that creative outcomes emerge out of the association or combination of various elements. Corazza and Agnoli (this volume) make the same point about science and engineering while Botella (this volume) stresses the role of combination in graphic art. And yet, while the way connections or analogies are established is closely scrutinised, questions about the origin and temporal dynamic of creative combinations are almost absent from the literature. These questions point us to the cultural history of the person (ontogenesis) in relation to that of society (sociogenesis). This is because the 'elements' being combined are not the sole product of isolated minds but represent ideas or types of knowledge acquired through socialisation, material action and communication with others.

Analogies connect two (or more) domains but the very nature of domains is sociocultural, not personal. Moreover, in design, art and science, analogies, and combinations are never made entirely 'in the mind'; creators use a variety of material and technical tools to access, explore, and materialise the outcomes of their combinatorial work.

A second example is offered by *perspective-taking*. Buisine and Bourgeois-Bougrine (this volume) make a clear case in their chapter about the importance of taking the perspective of users in engineering design. Following the established philosophy of the Personas method, whereby creators use different psychological profiles to approximate the perspective and needs of different people, the two authors fostered students' capacity for perspective-taking. The outcomes were satisfactory, although the authors noted that, instead of a Need-Seeker approach, their students followed more of a Market-Reader one, holding less creative potential. What they discovered to be useful, however, is employing off-target user profiles in order to tap into basic and unmet needs at the start of the creative process. They concluded that increasing the variety of perspectives by building multidisciplinary work groups would be highly beneficial for creativity in design. Similarly, Bourgeois-Bougrine and Glăveanu (this volume) noted the role played by empathy and perspective-taking in character construction in the case of screenplay writers. First and foremost, writers need to understand the 'heart' of the project by discussing it extensively with directors, producers and, if it is the case, co-writers. It is only by seeing the project from the perspective of different participants involved that stories are constructed and characters built. Afterwards, imagining the history, background, and perspectives of different characters is essential for defining dialogues and scenes. This type of identification gives direction to the project but also increases the degree of flexibility. Botella (this volume) mentions in this regard that, in graphic art, being creative means being flexible or able to analyse a problem from different angles and move between them. Flexibility is considered though as a 'cognitive factor' and this label overlooks the sociomaterial complexity of perspective-taking. In fact, to take or rather make a perspective (as nobody can literally 'take' the perspective of someone else) is a cultural achievement that involves social interaction, symbolic mediation, and position exchange

(Gillespie & Martin, 2014). While perspectives are mainly constructed in an imaginative manner, imagining what others experience has a developmental history marked by embodied changes of position in pretend play. Moreover, perspective-taking and empathy would be impossible outside the use of symbols that allow us to take distance from our concrete experience of the world and envision the experience of others. Taking perspectives is thus not only a mental process but a socially-enabled, culturally-mediated form of action essential for creative expression across domains (see Glăveanu, 2015).

A third example is represented by Corazza and Agnoli's (this volume) reference to the ability to *produce hypotheses* as central to creativity in science and engineering. The two authors note, in this context, the effort of interpreting what is observed as a constant task for scientists or engineers working on ill-defined problems. This ability can, as well, be translated from cognitive to sociocultural terms by referring to the uniquely human quest for meaning and the future-oriented nature of human activity (Valsiner, 2013). These two characteristics are key for creative work which, in essence, requires creators to give new meaning to existing or surprising phenomena and construct explanations for them. Meaning-making is at once cognitive and cultural because it involves the manipulation of signs and symbols formed and transformed within social interaction. To formulate a hypothesis, creators need to take into account both what is present in the here-and-now and, at the same time, use symbolic means, to take distance from it in order to explore the absent and the possible (Zittoun & Glăveanu, 2018). This capacity to imagine is equally personal and sociocultural because even the most idiosyncratic images or explanations conjured by our imagination build on experience and the resources made available by a shared culture.

Finally, the classic personality trait *openness to experience*, so often associated with creativity (at least since McCrae, 1987), goes well beyond an individual-level variable. Piirto (this volume) includes openness to experience among her five core attitudes for creators. She notes that creative writers tend to be especially open in trying out new things and changing their point of view. They explore a variety of spaces, engage many people, and explore different ways of doing things. From a sociocultural perspective, they are, in other words, open to difference (Glăveanu & Beghetto, 2017). Difference is the fundamental condition of living in the world and a primary source of creativity—the difference between self and others, objects and symbols, between past, present, and future, among others (Glăveanu & Gillespie, 2015). Faced with difference, we can either ignore or deny it or try to creatively engage with it by either bridging differences or, on the contrary, widening them. In the sciences, the former is often the case when, for instance, new creative ways are found to bridge the gap between our models of the world and reality. In art, we encounter frequently a different process at work: widening difference by problematising what is known or taken-for-granted. Either way, openness to difference constitutes a prerequisite for creativity and, I would argue, a more meaningful way of understanding the phenomena typically grouped under openness to experience.

In the end, what is the 'extended' creative mind? It is a view of psychology and creativity that goes explicitly beyond cognition or, rather, that tries to 'socialise' cognition and recognise it as culturally (i.e., socially, materially and symbolically) embedded. For too long we have pointed towards the head as the locus of creativity, imagining light bulbs miraculously being lit or little wheels turning on and on inside the mind or the brain. These little wheels bear different names in the traditional literature, from analogical thinking and perspective-taking to openness to experience, and so on. What would it mean to place these wheels, though, not in the restricted space between our ears but in a much wider space of society and culture? In other words, to see them engaging not only ideas but objects, other people, and institutions? This is the essence of an extended mind approach to creativity. However, in order to fully grasp its consequences, we need to see this mind in action, an issue I go on to discuss in the next section.

Creativity as Situated Action

As I argued before, to consider creativity as a form of action or activity means to recognise its psychological, social, material, and ultimately cultural nature. Human action is necessarily cultural: it is goal-oriented, symbolically mediated, and best understood as inter-action (between people, people and objects, etc.). Creativity in this context is both a quality of our action (Joas, 1996) and designates a form of activity in itself, in those contexts in which being creative is one of the main goals of doing something. At a more general, societal level, we label as creative activity all those actions taking place within cultural domains considered by definition 'creative' (e.g., the arts), although this broad label serves as much to promote as to exclude people from participating in creativity. As such, I define here creative action as action leading to the generation of outcomes or processes considered new, original, and meaningful or significant by creators and/or other people (see also Stein, 1953).

To study creativity in action or activity, we need to consider, again, the psychological, social, and material dimensions of creating. In other words, we need to pay attention to the *distribution of creativity* along different lines, from social to material and temporal (Glăveanu, 2014). Traditionally, in the study of the creative process, the psychological and temporal came to the fore at the expense of sociomaterial elements. Nonetheless, a sociocultural analysis of creative action broadens the scope by focusing on obstacles and facilitators, stages of the activity, social and material relations, as well as various outcomes, from artefacts to affective states (Glăveanu et al., 2013). More than this, it pays particular attention to the contextual nature of activity as no two creative actions, even when performed by the same person, are identical. Human activity is, in this sense, situated, meaning that it cannot be carried out or understood outside of its context. This observation contributes to the domain generality—domain specificity debate within creativity studies with a strong argument for the latter. As Kaufman and Baer (2005) describe in their amusement park theory of creativity (ATP), there are both similarities and differences between creative actions in different domains. Even when similarities are deduced analytically, by considering how close domains are from each other (e.g., if they share the same thematic area or not), the experience of creating is unique at the level of each task we engage in. The chapters included in this volume give us a good understanding of this 'mosaic' of differences and overlaps.

For example, Bonnardel, Wojtczuk, Gilles and Mazon (this volume) outline three main macro-processes or stages in design activities: (a) definition and redefinition of the creative problem, (b) openness to aesthetic dimensions, new experiences, and ideas, and (c) self-assessment or reflexive evaluation. As such, designers are seen as moving in their activity between defining problems, generating solutions, and testing them within an iterative or recursive process. Similarities can be found with engineers. Buisine and Bourgeois-Bougrine (this volume) discuss the new product design process in terms of: (a) translation of needs, (b) interpretation of needs, (c) product definition, and (d) product validation. The focus on needs is unique but the alternation between definition and validation resonates across the book. In science, Corazza and Agnoli (this volume) offer the simplest description of the creative process in terms of (a) inception, (b) ideation, and (c) impact. In this model, different actions serve different purposes and it is the main goal that differentiates between stages in scientific activities. Botella (this volume) includes many more steps in her discussion of the creative process in graphic art. Artists, she explains, start from an idea or 'vision', they then go through documentation and reflection, then prepare the first sketches and prototypes, do preliminary work, test and remake their products and, at times, prepare a series based on these experiments. While rich in detail, this depiction of the creative process returns to some previously mentioned mechanisms: ideation and reflection, working and reworking the product, etc. Finally, Bourgeois-Bougrine and Glăveanu (this mention) simplify the creative process in screenplay writing by referring to basically three phases: (a) impregnation, (b) planning for action/structuring, and (c) production.

What can we notice from the above when it comes to creative activity? First of all that it is often cyclical, involving ideation and evaluation, production and testing, writing and rewriting, etc. Second, these stages can be found in different domains, sometimes domains from different thematic areas, while they never involve precisely the same kind of actions and sets of constraints. Both scientists and artists experiment in their creative work but they do so in markedly different ways. Last but not least, the social and material environment participates in creative action in different ways. The technologies used in music and science are different, the social system of evaluation sets the arts and engineering apart, the pressure to respect constraints varies between scientists and writers, for instance. In other words, creative action is culturally situated and can only be understood in relation to its context (see also Barbot & Webster, this volume). It is all the more surprising then to notice the fact that many studies in the psychology of creativity report their findings as if they applied to creativity in general, instead of coming out of particular tasks (e.g., collage) within particular domains (e.g., art). This goes beyond a simple omission and constitutes an academic practice that reinforces the idea of a universal creative process and minimises the role of context and culture. And yet, there is no 'creativity in general' and, I would argue, no general creative potential (equivalent to the g factor in intelligence) conceivable outside of domains of activity and sociocultural practices. The simple reason is that, as a quality of human action, creativity cannot be separated from the means, goals, and constraints of action.

In the end, the sociocultural approach doesn't deny similarities in creative action and it does pay attention to general or generalisable features of creative work across domains. As such, it is not mindlessly missing the forest because of a concern for each and every branch within it. If abstracting features and building generalisable frameworks are all characteristic for science, then the approach proposed here makes no exception in this regard. What it doesn't support, however, is the quest for a unique, universal, and general process of creativity that transcends context and sidesteps culture. One candidate for such a process seems to be divergent thinking, the topic of many psychometric and experimental efforts in the psychology of creativity, one that is often confused with creativity as a whole. What would a sociocultural framework focused on similarities across domains propose as an alternative?

The Cultural 'Movements' of Creativity

In this final section I would like to develop a basic sociocultural model of the creative process based on what I call the *dual movement* of creativity. I prefer the notion of movement to those of step, stage, phase, or even process because it avoids the dangers of reifying or making linear a phenomenon that, as mentioned above, is basically cyclical or recursive. Also the idea of movement is much closer to that of action and it opens up new and interesting theoretical horizons, reflected for example in the study of rhythms (Glăveanu, 2016) and pathways of creating (Tanggaard & Beghetto, 2015). What are the two basic movements of creative activity? Immersion and detachment. In other words, being fully engaged in action, experiencing it first-hand and becoming absorbed in its sensorial qualities (something Csikszentmihalyi, 1990, referred to as flow) and, at the same time, being able to take a step back, put some distance between self and work, and reflect on it often from a new perspective. This is a paradoxical state I theorised recently as 'immersed detachment' (see Glăveanu, 2018). The dual-movement framework of creativity is equally psychological, embodied/material, and social. It is by definition a cultural model, as the capacity to detach and reconnect with the here-and-now of our experience is necessarily scaffolded by culture (Zittoun & Gillespie, 2016).

Traces of these two movements can be easily found in descriptions of the creative process across the book: the openness to new ideas and the reflexive evaluation of these ideas mentioned by Bonnardel et al. (this volume), product definition and product validation (Buisine & Bourgeois-Bougrine, this volume), making and testing sketches or prototypes (Botella, this volume), writing, reading, and rewriting scenes (Bourgeois-Bougrine & Glăveanu, this volume), unconscious and conscious exploration (Toulouse, this volume). It also resonates widely with the well-known Geneplore model in creative cognition studies (Finke, Ward, & Smith, 1992). Indeed, immersion is often associated with idea generation, and detachment is often achieved by adopting an evaluative stance on the ideas that had just been generated. But the dual movement referred to here cannot be reduced to idea generation and idea evaluation. When immersed in one's activity, creators don't just come up with ideas but experience the work first hand, act on their impulses, enjoy or become tense because of it, and sometimes let the object take over the process and let themselves be guided by the developing artefact (an experience that is very common in art, writing, but not absent from science or design). Conversely, when detached from their work, creators not only assess it but notice new constraints, new affordances, reflect on how well it represents their initial ideas and on what needs to be done next. The dynamic between immersion and detachment mirrors closely what John Dewey (1934) theorised as the cycle between doing and undergoing in creative activity, between acting on the world and experiencing the reaction of the world to what is being done. It is a dual movement that can be used to describe both macro-stages of creating in different domains (when, for instance, some days or even months are dedicated to constructing something while others to evaluating or testing the creation) and the micro-processes that take place in moment-to-moment action (for example, painters applying paint and then stepping back to observe the outcome of their action; Glăveanu, 2015).

This alternation between immersion and detachment, and sometimes their simultaneity, are often referenced in the chapters included here, albeit in different terms. Martín et al. (this volume) talk about two mental models in composition: the creative mode and the evaluation mode. In the former, composers find new solutions to their problems whereas in the latter they revisit their ideas and correct them if necessary. Switching between these modes is considered by the authors as one of the most challenging aspects of music composition and creation in general. I agree with this statement although I wouldn't reduce these modes to 'mental' models but ground them in action, as movements. Nor would I label only one of them as 'creative'; in fact it is precisely their interplay that gives birth to new and meaningful outcomes. Bourgeois-Bougrine and Glăveanu (this volume) report in their chapter the phenomenological experience of screenplay writers who perceive their activity as a cycle of continuously generating and testing alternatives. In a metaphorical way, one of their participants makes a strong case for the value of detachment in creative work when saying that, just like in the case of painting, you need to be patient and 'let things dry'. Equally, scripts need periods of letting go and returning to them with a fresh eye, otherwise doing too much can turn everything muddy. It is therefore not only a cognitive detachment that we should look for but a physical and, most of all, affective detachment that increases one's chances of success. Some writers even use automatic writing (see Piirto, this volume) in their immersion moments in order to capitalise on the non-reflexive and non-judgemental freedom afforded by immersion. In design, both Bonnardel et al. (this volume) and Nelson (this volume), emphasise the iterative process of 'seeing-moving-seeing', creating mental representations, trying to apply them in practice in the making of objects, and adjusting both representations and objects as the work progresses. The opportunistic processes at play are often experienced as a dialogue between designer and object, a dialogue that, I would add, goes through social others (e.g., users, the person who commissioned the work, and so on). In science, the story of penicillin, referred to by Corazza and Agnoli (this volume) illustrates well the dual movement and the role of detachment for turning what was basically an accident into a scientific discovery. Had Fleming not been able to take distance from the situation and see the culture of staphylococci contaminated by a fungus as something else than a 'mistake', we would have waited longer for a crucial advance in medicine to take place.

In summary, the immersion–detachment framework of the creative process is widely applicable and yet, at all times, needs to be considered in a contextual manner. Not all creators and not all domains experience immersion or detachment in the same way and, in each case, a variety of psychological, social, and material factors come into play. The cultural basis of the creative process is emphasised here by the fact that creativity is assumed to involve a form of *repositioning* of the creator towards the situation. This repositioning can be physical, social or symbolic/imaginative, or all these at the same time. But the essential aspect of it is that it allows creators to alternate between a 'first-person' (immersed) perspective on the emerging artefact and a 'third-person' (detached) perspective—i.e., seeing oneself, the situation, or the work as another person would (Mead, 1964). Elsewhere I theorised this dynamic as the *perspectival model* of creative action (Glăveanu, 2015). For the purposes of this chapter it suffices to say that the essence of creativity relates to placing different perspectives in dialogue, a process enabled by the alternating movements of immersion and detachment discussed here. With the means of culture, both physical and symbolic, we are capable of experiencing the world from different positions and building different perspectives on it. The key challenge for creativity researchers and especially for practitioners is to understand, cultivate, and guide this capacity by noticing how, when, and why creators move between immersion and detachment, and helping them reflect on these movements.

Creativity, Mind and Culture

I started this chapter by reflecting on the fact that culture is often ignored or marginalised in studies of creativity and ended it with a sociocultural proposal for the creative process, one that considers the psychological aspects of creating in relation to the sociomaterial world creators live and work in. Undoubtedly the picture of creativity is extremely complex and those models that exclude the social and the cultural often do so in an effort to become parsimonious (Runco, 2015), to focus on what is 'essential' for the phenomenon and nothing else. The big problem with such models is that they turn easily from simple to simplistic and, in an effort to streamline the creative process, they might end up doing away with creativity itself (and replacing it with something more basic, for example ideation, a common tendency in neuropsychological studies of 'creativity'). Other models try to be more comprehensive by expanding our view from cognitive to conative, emotional, and environmental factors (see the multivariate approach, Lubart, Mouchiroud, Tordjman, & Zenasni, 2003). And yet these models are more structural than dynamic and tell us relatively little about the creative process and how it mobilises factors at different moments— one of the many concerns of the chapters included in this book, see for example Botella (this volume).

What I would like to push for in the end is an even deeper integration of culture, what we usually call 'environment', in creativity theory. My aim is to show that rather than being what prompts, facilitates, supports or shapes the process from the outside, culture is in fact part and parcel of creative processes and, ultimately, it makes them possible. The dual-movement framework of immersion and detachment represents one such attempt at developing a sociocultural *as well as* psychological model of creative action.

There are many theoretical and practical consequences associated with this new and expanded framework as it ultimately challenges our methods and practices as creativity researchers. But one of the consequences that is often forgotten and I would like to underline in the end is the fact that it is not only the case that culture 'bridges' mind and creativity—in turn, the sociocultural study of creativity sheds new light on the highly complex relation between mind and culture. To be a creative agent means much more, according to this paradigm, than to think divergently or to combine information in ways not seen before. It means to be capable of not only appropriating but also *participating* in culture. Of course, as I hope it became clear in this chapter, I am not referring here to culture as only the 'high culture' of museums or science laboratories but in a much wider sense, as the shared environment of ideas, practices, and material arrangements in which we are immersed. Traditional views of socialisation and, more specifically, schooling, emphasise the role of society in reproducing the same structures and knowledge in the process of sharping individual minds and actions. To be a creative person means, first and foremost, to be capable of detachment and reflection, of using cultural elements in order to understand oneself and co-construct culture. This is the reason why creativity should feature high on the agenda of sociocultural thinkers, independent of whether they come from psychology or other disciplines (Glăveanu, Gillespie, & Valsiner, 2015). Barbot and Webster (this volume) draw our attention to the need of theorising agency in creativity research in ways that are not individualistic but bridge the person–culture divide. A sociocultural theory of the creative process sets this for itself as a higher goal, one that requires further inter and multidisciplinary dialogue as well as a deeper and more meaningful dialogue between theory and practice. Thus, in many ways, the work we have in front of us has just begun.

References

Baer, J., & Kaufman, J. C. (2005). Bridging generality and specificity: The Amusement Park Theoretical (APT) model of creativity. *Roeper Review, 27,* 158–163.

Barbot, B., & Webster, P. R. (this volume). Creative thinking in music: Processes of children and adolescents. In T. I. Lubart (Ed.), *The creative process: Perspectives from multiple domains* (pp. 255–273). London: Palgrave.

Bonnardel, N. (2000). Towards understanding and supporting creativity in design: Analogies in a constrained cognitive environment. *Knowledge-Based Systems, 13,* 505–513.

Bonnardel, N., Wojtczuk, A., Gilles, P.-Y., & Mazon, S. (this volume). The creative process in design. In T. I. Lubart (Ed.), *The creative process: Perspectives from multiple domains* (pp. 209–227). London: Palgrave.

Botella, M. (this volume). The creative process in graphic art. In T. I. Lubart (Ed.), *The creative process: Perspectives from multiple domains* (pp. 59–87). London: Palgrave.

Botella, M., Nelson, J., & Zenasni, F. (2016). Les macro et microprocessus créatifs [Macro and micro creative processes]. In I. Capron-Puozzo (Ed.), *La créativité en éducation et en formation. Perspectives théoriques et pratiques* [Creativity in education and formation. Theoretical and practical perspectives] (pp. 31–44). Paris: Albin Michel.

Bourgeois-Bougrine, S., & Glăveanu, V. P. (this volume). Collaborative Scriptwriting: Social and Psychological Factors. In T. I. Lubart (Ed.), *The creative process: Perspectives from multiple domains* (pp. 123–154). London: Palgrave.

Buisine, S., & Bourgeois-Bougrine, S. (this volume). The creative process in engineering: Teaching innovation to engineering students. In T. I. Lubart (Ed.), *The creative process: Perspectives from multiple domains* (pp. 181–207). London: Palgrave.

Clark, A., & Chalmers, D. (1998). The extended mind. *Analysis, 58*(1), 7–19.

Corazza, G. E., & Agnoli, S. (this volume). The creative process in science and engineering. In T. I. Lubart (Ed.), *The creative process: Perspectives from multiple domains* (pp. 155–180). London: Palgrave.

Csikszentmihalyi, M. (1990). *Flow: The psychology of optimal experience.* New York, NY: HarperCollins.

Csikszentmihalyi, M. (1998). Creativity and genius: A systems perspective. In A. Steptoe (Ed.), *Genius and the mind: Studies of creativity and temperament* (pp. 39–66). Oxford: Oxford University Press.

Dewey, J. (1934). *Art as experience.* New York, NY: Penguin.

Finke, R. A., Ward, T. B., & Smith, S. S. (1992). *Creative cognition: Theory, research, and applications.* Cambridge, MA: MIT Press.

Gillespie, A., & Martin, J. (2014). Position exchange theory: A socio-material basis for discursive and psychological positioning. *New Ideas in Psychology, 32,* 73–79.

Glăveanu, V. P. (2011). Creativity as cultural participation. *Journal for the Theory of Social Behaviour, 41*(1), 48–67.

Glăveanu, V. P. (2013). From dichotomous to relational thinking in the psychology of creativity: A review of great debates. *Creativity and Leisure: An Intercultural and Cross-disciplinary Journal, 1*(2), 83–96.

Glăveanu, V. P. (2014). *Distributed creativity: Thinking outside the box of the creative individual.* Cham: Springer.

Glăveanu, V. P. (2015). Creativity as a sociocultural act. *Journal of Creative Behavior, 49*(3), 165–180.

Glăveanu, V. P. (2016). Rhythm. In V. P. Glăveanu, L. Tanggaard, & C. Wegener (Eds.), *Creativity: A new vocabulary* (pp. 129–136). London: Palgrave.

Glăveanu, V. P. (2018). Epilogue: Creativity as immersed detachment. *Journal of Creative Behavior.* Online first.

Glăveanu, V. P., & Beghetto, R. A. (2017). The difference that makes a 'creative' difference in education. In R. A. Beghetto & B. Sriraman (Eds.), *Creative contradictions in education* (pp. 37–54). New York, NY: Springer.

Glăveanu, V. P., & Gillespie, A. (2015). Creativity out of difference: Theorising the semiotic, social and temporal origin of creative acts. In V. P. Glăveanu, A. Gillespie, & J. Valsiner (Eds.), *Rethinking creativity: Contributions from social and cultural psychology* (pp. 1–15). Hove/New York: Routledge.

Glăveanu, V. P., & Lubart, T. (2018). Cultural differences in creative professional domains. In A. K.-y. Leung, L. Y.-Y. Kwan, & S. Liou (Eds.), *Handbook of culture and creativity: Basic processes and applied innovations* (pp. 123–141). Oxford: Oxford University Press.

Glăveanu, V. P., Gillespie, A., & Valsiner, J. (Eds.). (2015). *Rethinking creativity: Perspectives from cultural psychology.* London: Routledge.

Glăveanu, V. P., Lubart, T., Bonnardel, N., Botella, M., de Biaisi, M.-P., Desainte-Catherine, M., et al. (2013). Creativity as action: Findings from five creative domains. *Frontiers in Educational Psychology, 4,* 1–14.

Hutchins, E. (1995). *Cognition in the wild.* Cambridge, MA: MIT Press.

Joas, H. (1996). *The creativity of action.* Cambridge: Polity Press.

Lubart, T. I. (2000). Models of the creative process: Past, present and future. *Creativity Research Journal, 13,* 295–308.

Lubart, T. I., Mouchiroud, C., Tordjman, S., & Zenasni, F. (2003). *Psychologie de la créativité* [Psychology of creativity]. Paris: Armand Collin.

Martín, D., Pachet, F., & Frantz, B. (this volume). The creative process in lead sheet composition. In T. I. Lubart (Ed.), *The creative process: Perspectives from multiple domains* (pp. 275–296). London: Palgrave.

McCrae, R. R. (1987). Creativity, divergent thinking, and openness to experience. *Journal of Personality and Social Psychology, 52*(6), 1258–1265.

Mead, G. H. (1964). *Selected writings: George Herbert Mead*, ed. A. J. Reck. Chicago: University of Chicago Press.

Moran, S., & John-Steiner, V. (2003). Creativity in the making: Vygotsky's contemporary contribution to the dialectic of development and creativity. In R. K. Sawyer, et al. (Eds.), *Creativity and development* (pp. 61–90). Oxford: Oxford University Press.

Nelson, J. (this volume). Modelling the creative process in design: A sociocognitive approach. In T. I. Lubart (Ed.), *The creative process: Perspectives from multiple domains* (pp. 209–227). London: Palgrave.

Piirto, J. (this volume). The creative process in writers. In T. I. Lubart (Ed.), *The creative process: Perspectives from multiple domains* (pp. 89–121). London: Palgrave.

Runco, M. A. (2015). A Commentary on the social perspective on creativity. *Creativity. Theories—Research—Applications*, *2*(1), 21–31.

Shweder, R. (1990). Cultural psychology—What is it? In J. Stigler, R. Shweder, & G. Herdt (Eds.), *Cultural psychology: Essays on comparative human development* (pp. 1–43). Cambridge: Cambridge University Press.

Stein, M. (1953). Creativity and culture. *Journal of Psychology*, *36*, 311–322.

Sternberg, R. J., & Lubart, T. I. (1995). *Defying the crowd: Cultivating creativity in a culture of conformity*. New York, NY: Free Press.

Tanggaard, L., & Beghetto, R. (2015). Ideational pathways: Toward a new approach of studying the life of ideas. *Creativity: Theories—Research—Applications*, *2*(2), 130–144.

Toulouse, I. (this volume). The importance of being earne(arti)st. In T. I. Lubart (Ed.), *The creative process: Perspectives from multiple domains* (pp. 19–27). London: Palgrave.

Valsiner, J. (2013). *An invitation to cultural psychology*. New Delhi: Sage.

Zittoun, T., & Gillespie, A. (2016). *Imagination in human and cultural development*. London: Routledge.

Zittoun, T., & Glăveanu, V. P. (Eds.). (2018). *The Oxford handbook of imagination and culture*. Oxford: Oxford University Press.

Index

© The Editor(s) (if applicable) and The Author(s) 2018
T. Lubart (ed.), *The Creative Process*, Palgrave Studies in Creativity and Culture,
https://doi.org/10.1057/978-1-137-50563-7

CPI Antony Rowe
Eastbourne, UK
December 01, 2019